# Regression with Microdata and Microsoft Excel®

Humberto Barreto
Professor of Business Analytics
DePauw University
2026

This book was typeset in $\LaTeX$ with various packages in TeXstudio. I was helped repeatedly by resources at tex.stackexchange.com. I am awed by this software and its community. I offer a deep bow to those who made these tools freely available and continue to provide support.

Humberto Barreto
April 16, 2026

ACKNOWLEDGMENTS

While he did not work with me on this book, these materials bear the imprint of my long-time colleague, Frank Howland.

This work was financially supported by Hewlett-Melon funds at DePauw University.

Many cohorts of students in ECON 385 and BUS 305 at DePauw University gave me valuable and constructive feedback. I hope this book helps future students, at DePauw and everywhere around the world.

Thank you all.

**To my best friend, Tami.**

**I am forever grateful.**

# Contents

# Preface

Figure 1: Carl Friedrich Gauss (1777-1855) generated by ChatGPT.

In math, many consider Gauss (rhymes with house) to be the GOAT (greatest of all time). This book often uses the normal curve, which is more technically known as the Guassian distribution. You will also learn what the Gauss-Markov Theorem says and why it is so important. He made major discoveries not just in statistics, but in a wide range of mathematics and physics.

What would Gauss think of this book? I am afraid he would not be a fan. It is way too slow. And wordy. He would wonder why there are so few equations and formulas. I bet Gauss could put all of the content in this book on a single page!

Of course, then only experts like him would be able to understand it. I am not focused on Gauss and people like him. There are lots of books for them and they are going to figure it all out anyway. I am trying to reach the novice, the math phobic, and those who think they will never understand sophisticated quantitative ideas.

I am assuming you have taken a statistics course and that you are in your second or third year of college. You have heard of the SD, but you might not be positive about what it means. The same is true of hypothesis testing.

My teaching strategy is to make everything ***concrete*** and ***visual***. Consider this quote from an excellent, conventional econometrics book:

> Imagine that we repeatedly drew samples of size $n$ from the population (with replacement), recorded the sample mean and sample variance of each, wrote those on a clean chip, and put the chip into a bowl. As this process was continued indefinitely, the proportions in the new bowl would stabilize, the relative frequencies approaching the joint probability distribution of sample means and sample variances in size-$n$ random sampling from the original population. Such a derived distribution—a probability distribution of one or more sample statistics for a given sample size—is called a *sampling distribution.* Goldberger, A. (1998) *Introductory Econometrics*, p. 44.

If you are like the vast majority of people, you have no idea what this says. The jargon is a barrier, but so is the very first word: *imagine.* Mathematicians are great at abstract thinking, but this is not a good way to teach. Too many resources are devoted to mental construction of abstract ideas.

This book will not ask you to imagine anything. It will use Excel to show every number and formula. It will use simulation to display what is going on. You will be able to see the results. You will truly understand complicated concepts like the sampling distribution. There is no doubt that the material is quite challenging, but this is an excellent opportunity to master these ideas.

This is not a typical math book. You will not be memorizing formulas and the focus is on understanding concepts. This approach has worked for many students and I hope it works for you!

Humberto Barreto
hbarreto@depauw.edu
Greencastle, Indiana

# User Guide

This book is a manual for actively working with and manipulating content in Excel. There are, however, many versions of Excel—which one is needed?

Use ***Windows*** (not Mac) ***desktop*** (not browser) ***Excel***.

Some of the work can be done on a Mac, but the best solution for Mac users is to emulate Windows with software such as Parallels or Boot Camp. For students at an educational institution, accessing a Windows machine from a server is an easy solution for Mac users. Students at my university can login at desktops.depauw.edu from a Mac, Chromebook, or tablet (anything that has a browser) to access a Windows machine with desktop Excel (and other Office applications).

Online versions of Excel do not support Visual Basic and other advanced features. Thus, you cannot get full functionality using Excel within a web browser. Do not use Excel in a browser to do the work in this book.

In addition to Windows desktop Excel, this book presumes that you have seen a spreadsheet before. You can open Excel and write a formula that adds cells together. As you will see, however, Excel is much more than a simple adding machine. You will learn how to use Excel in much more advanced ways. In addition to analyzing data and learning many new Excel functions, you will learn how to use special Excel files called *add-ins* that solve optimization problems and run simulations.

The most important thing you can do as you read this book is **experiment**. You might find yourself wondering, "What would happen if this cell was 10 instead of 1?" Do not just wonder, change the cell and see what happens! There is deep neuroscience at work here. When you are in control and making up your own questions, you learn best. The beauty of this approach is that everything is alive and you can make points move and lines shift. Take full advantage. Be engaged.

*EXCEL TIP* Never settle for passive reading. Always do the steps and experiment as you work, changing things to see what happens.

If something is not working the way you expect, there are many possible causes. It is always a good first step to close Excel completely and reopen it. Even if this does not fix the problem, slowly repeating the steps will help you debug or describe what is happening.

Finally, consider using generative AI (such as ChatGPT, Gemini, or Claude) if you get stuck. You can even upload an Excel file to help figure out what is going on.

## Guiding Principles

This book is different from an Excel-based book I co-authored two decades ago (Barreto and Howland, *Introductory Econometrics*, Cambridge University Press, 2006). Instead of working with prepared Excel files, each lesson starts with a blank spreadsheet and you construct everything. This book also is much more focused on fundamental concepts and leaves out more advanced material that is not essential.

Each chapter in this book has material for one day in an hour-long class. It makes sense to work your way from the first to the last chapter, in order.

Although not required, it is better to work with a printed version so that you can write notes and check off completed steps. If you choose to work with an electronic version, two screens (or at least a large monitor) is recommended. Use your PDF software to write notes and highlight text.

## Website

This book is open access and is distributed through many outlets. Visit and bookmark the book's website at dub.sh/regcourse.

This site has the latest version of the book, along with all of the add-ins and finished files. You should build your own Excel workbooks, but you can always check your work by downloading files and comparing them to yours.

# 1. The Journey Begins

Our research question is complicated and hard to answer: *What is the rate of return to education?* In other words, we know people with college degrees make more than high school graduates, but exactly how much more? We need a number because we also want to know if the additional income is enough to make college worth it. This question about the return to education is usually answered with regression analysis. Thus, it is a perfect vehicle for learning about regression.

Our journey begins with the income people get from working. We will get data on earned income and do some basic statistical analyses. Later, we will explore the relationship between earned income and education and other variables. We start slowly and carefully and gradually pick up steam.

## 1.1 Getting IPUMS CPS Data

The Current Population Survey (CPS) is a monthly household survey run by the U.S. Census Bureau and the Bureau of Labor Statistics.

> Initiated in the 1940s in the wake of the Great Depression, the survey was designed to measure unemployment. A battery of labor force and demographic questions, known as the "basic monthly survey," is asked every month. Over time, supplemental inquiries on special topics have been added for particular months. Among these supplemental surveys, the Annual Social and Economic Supplement (hereafter referred to as the ASEC) is the most widely used by social scientists and policymakers. The ASEC along with basic monthly data from the CPS provides the data for IPUMS-CPS. (cps.ipums.org/cps/about.shtml)

IPUMS used to be an acronym for Integrated Public Use Microdata Series, but in 2016 it adopted IPUMS as its name. It is the world's largest individual-level population database. The *micro* in *microdata* means that each record (or row in the spreadsheet) is an observation (in this case, a person), with a numerical code for every characteristic. Microdata are usually "big data" sets with hundreds of thousands or more observations. IPUMS takes the CPS (and other massively complex surveys) and makes them accessible.

IPUMS gives a user-friendly interface to access the CPS. We pick a few variables, choose some options so Excel can read the data, and download the file.

STEP Create an account (click *Register*) by filling out an application at cps.ipums.org. Be sure to completely and honestly fill out the form. If you are a student, use your institution's email address.

You should get a response fairly quickly, but you may have to wait a little bit. You will receive an email when access is granted.

STEP Login at cps.ipums.org and click *Select Data*.

STEP Click *Select Samples* and select only the latest ASEC sample year. Check *Select All Samples* and then uncheck to remove all samples. Then check only the latest year. Click the *Basic Monthly* and make sure no samples are selected. Click *Submit Sample Selections*.

IPUMS shows that you have one sample selected in your cart. Now we get the variables we want from this sample. IPUMS organizes the variables logically so AGE and SEX are under *Person* and *Demographic*. Variables are also listed alphabetically (if you know the name, like EDUC, and you can search for a variable (the search term "income" gives INCWAGE).

Clicking on a variable name takes you to a page with information about the variable. There are several tabs that describe the variable and its codes. As you collect the variables below, be curious and explore a little.

STEP Find AGE, SEX, EDUC, and INCWAGE by using the three approaches mentioned above and clicking the + to add the variable to your cart.

STEP Click *View Cart* to see the variables you will be downloading. Confirm that you have AGE, SEX, EDUC, and INCWAGE (they do not have to be in that order) along with other preselected variables added by IPUMS.

IPUMS shows your *Extract Request*, with the data format and file size.

STEP Make sure the data format is CSV (comma-separated values is what spreadsheets, like Excel, use). Click *Change* and select *Comma delimited (csv)* if needed.

We want to compare the earnings of adult workers, so we will get data on people ages 25 to 64 years old. This gives them a few years to get out of college and many people retire at 65.

*STEP* Click *Select Cases* and restrict the age range to 25 to 64 years old.

You are ready to submit your selections and get the data.

*STEP* Click *Submit Extract* to request the data.

Refresh the screen a few times until you see *Download CSV* under the *Formatted Data* column. IPUMS will also send you an email with a link when your data are ready. The email includes citation text which you should include in the references section of any paper or presentation using this data source.

*STEP* Click *Download CSV* to get a zipped archive of the data.

You need software to uncompress the .gz file. Your computer may have this built-in or you can use www.7-zip.org which is free and open source. To unzip a file with 7-zip, right-click on it, select *7-zip*, select *Extract files...*, click the box on the top-right with three dots, navigate to an appropriate folder, and click *OK*. Other uncompression software will work in a similar fashion.

*STEP* Unzip the downloaded .gz file and open it in Excel.

## 1.2 Exploring INCWAGE in Excel

Begin by saving the .csv file as an Excel workbook.

*STEP* Click *File* and click *Save As*. Click *Save as type* and be sure to select *Excel workbook (*.xlsx)*. Name it well, describing what it contains, such as, *CPSIncomeEduc.xlsx*.

Usually, variables need to be recoded before they can be used, but INCWAGE is simply the reported income from work, such as salary, wages, and tips. We can begin working with INCWAGE right away. Likewise, AGE is simply how old the person said they were.

After the first row, which has the labels, each row in the Excel file is a person. It is easy to see the age of the first person (row 2) in your data set and their income, but the other variables require a codebook to tell us what the numbers mean.

We want to be organized and careful as we clean and prepare the data for analysis. Excel allows us to name individual sheets in a workbook. This will help us keep everything orderly and reproducible.

*STEP* Change the sheetname to *rawdata*.

Another important, but often omitted step, is to document what you do. Someone reading your workbook (maybe you a few weeks from now) needs a road map of how the variables were recoded and analyzed.

*STEP* Best practice is to add a sheet to your workbook called *Doc* where you will include all of the steps you take. It does not have to be detailed or even complete sentences, but should indicate how the raw data were cleaned and processed.

We begin preparing the data for analysis by removing people with insufficient earned income to qualify as full-time workers. The IPUMS documentation on INCWAGE tells us that this variable "reports each respondent's total pre-tax wage and salary income—that is, money received as an employee—for the previous year."

We can see, however, many zeroes. These are people who did not earn any income last year so they are not relevant to our question about how education affects earned income. We can sort the data to remove these people from our data set.

*STEP* Select any cell with data, click the *Data* tab in the ribbon, and click *Filter* (in the *Sort & Filter* group).

Excel prepares the sheet for sorting and filtering by putting little arrows in the cells in the first row. Notice that *Filter* is a toggle (you can see it is "pushed in"). You can click on it at any time to remove the arrows from the first row.

**STEP** Click the arrow in the INCWAGE cell and select the top choice, *Sort Smallest to Largest.*

Excel reorganizes the sheet, sorting the entire row (keeping the variables together for each observation). We can now delete everyone who had no earned income in the previous year. In fact, we will be even more aggressive in our definition of full-time worker.

But before we do this, it is worth pointing out that sorting is one of the most dangerous operations you can perform on a spreadsheet. It is remarkably easy to destroy a data set by sorting only part of it. Excel has improved over the years, with warnings about potential damage, but sorting remains a powerful and hazardous activity. Proceed with caution when you sort.

**STEP** Scroll down to see when the zeroes disappear and you will see they are followed by small earnings. Keep going until you reach 20,000.

Our arbitrary cutoff for a full-time worker is $20,000 based on making $10/hr for 2,000 hours in a year. You could make an argument for a different definition of full-time, but what really matters is that the decision is clear and transparent.

**STEP** Select the row (the entire row, by clicking on the row number) of the person making just under 20,000 and scroll back up to the top of the sheet, press and hold the *shift* key, and click row 2. With these rows highlighted, click *Delete* in the *Home* tab (in the *Cells* group).

We now have a data set with people ages 25 to 64 years old with a college degree who earned at least $20,000, which is our definition of full-time work. We need to update our *Doc* sheet before we continue.

**STEP** Add a note in your *Doc* sheet about this sort and delete procedure and save the workbook (keyboard shortcut ctrl-s). Saving regularly is good practice.

The 2024 ASEC has 50,124 observations (people) in the *rawdata* sheet. The minimum INCWAGE is 20,000 and the maximum (at the bottom of the sheet) is 1,399,999. If you downloaded the 2025 (or above) ASEC, you will have a different sample size and slightly different results in future work.

## 1.3 INCWAGE Visualization

Let's see how INCWAGE is distributed.

*STEP* Select the entire INCWAGE column and make a histogram chart by clicking the *Insert* tab and choosing *Histogram* from the Charts group. //
The histogram is very hard to read. It has a really long right tail. We can improve this chart.

*STEP* Double-click the $x$ axis and make the *Bin Width* 10000. Check the *Overflow Bin* and make it 400000. Make the title *INCWAGE*.

Now we can clearly see how income is distributed. It does have a long tail, but we can read that there are about 7,000 people in the sample making 20,000 to 30,000 dollars in income.

Income is not normally distributed. It does not have a bell shape, with most of the observations in the middle and fewer as you move away from the center. Instead, there are a lot of values at low income levels and then it gradually tapers off. This is how measures of earnings (such as wages and income) are usually distributed.

## 1.4 Median v Average

There are two commonly used measures of central tendency, the average and the median. Often, they agree. Is that true here?

*STEP* Use the Excel functions AVERAGE and MEDIAN in cells O1 and O2. What do you find?

In the 2024 ASEC, the median was $60,000. This means that half of the sample made less and half made more. The average, however, was almost $83,000.

The average is sensitive to large values and is being pulled up by the extremely high incomes. The median is unaffected by these values. It does not matter if the incomes above the middle value are barely higher or really higher.

The average and median of the numbers 1, 2, and 3 are both 2. Increase the 3 to 3 million and the average rockets. The median remains 2.

In fact, when the typical value of income is reported, it is often the median because it captures where most of the values are located.

However, there is something that average can do that the median cannot. And this property is also shared by regression. Let's see what this mysterious idea is all about.

Suppose you were asked to predict the income of a randomly chosen person in the sample. Suppose further that we would square the amount you missed by. Your goal is to find the best predictor in the sense of having the smallest squared deviation.

We can be more concrete. Suppose you guess the median, $60,000 and the random income turns out be $50,000. Then your score is actual minus predicted squared, so $50,000 - $60,000 = - $10,000, and squared it is $100,000. Notice that the squaring eliminated the minus sign and also that squaring penalizes missing since the deviation gets amplified.

Instead of a single random draw, we compute what the penalty would be for every observation and sum them. The difference between actual and predicted for each observation is called a residual and the best prediction is measured by the one that minimizes the sum of squared residuals.

*STEP* In cell Q1 enter the label *Predicted.* In cell Q2 enter the number 60000. In cell R1 enter the label *Residual* and in cell R2 enter the formula = *N2 - $Q$2* (assuming INCWAGE is in column N, if not, adjust the formula to use its column). Fill down.

Residuals are always actual minus predicted and they can be positive, negative, or zero. A positive residual means the guess was too low, while a negative residual occurs when the predicted value is too high. A perfect guess occurs when the residual is zero and that is what we want.

*STEP* In cell S1 enter the label *Squared Residual* and in cell S2 enter the formula = *R2^2.* Fill down. In cell T1 enter the label *Sum of Squared Residuals* and in cell T2 enter the formula = *SUM(S:S).*

Excel is displaying a large number formatted in scientific notation. We seek the predicted value that minimizes the *Sum of Squared Residuals*, which everyone calls *SSR*.

*STEP* Change cell Q2 to 70000. You can immediately see cell T2 change. Try a small value like 25000 and see how the SSR goes up. The SSR is also high when you set cell Q2 to a large guess like 150000.

The SSR is a function that depends on the predicted value. We can see its shape by using Excel's *Data Table*.

*STEP* In cell Y2 (leave cell Y1 blank) enter the number 50000. Enter 60000 in cell Y3, select cells Y2 and Y3, then fill down to 120000. In cell Z1 enter the formula $= T2$.

We are ready to insert the Data Table. It will compute the SSR for each value in column Y.

*STEP* Select cell range Y1:Z9, click the *Data* tab, click the *What-If Analysis* button (in the *Forecast* group) and select *Data Table*. Click in the column input box and click cell Q2 (where you have the predicted value). Click OK.

Excel inserts an array function, TABLE, in cells Z2 to Z9 that displays the SSR for the values in column Y.

*STEP* Select cells Y2:Z9 and make a Scatter chart.

Your chart clearly shows that the SSR function is U-shaped. The best prediction is the minimum of that function. But where exactly is it? We can use Excel's Solver to find it. Solver is an optimization algorithm. It tests many trial solutions very quickly, converging to an answer.

*STEP* Click the *Data* tab, then Solver (in the Analysis group) to bring up the Solver Parameters dialog box. If Solver is not available, then use the Add-in Manager to install it. Use Excel's Help if you are having trouble or visit support.office.com.

*EXCEL TIP* Press *alt, t, i* to access the Add-in Manager. Press the *alt* key, then release it and press the letter *t*, then release it and press the letter *i* to quickly bring up the Add-in Manager.

The Solver Parameters dialog box is initially empty. You need to give it the appropriate information before asking it to search for a solution. It always needs specific cell addresses for the Objective and Changing Variable Cells inputs. Some problems are constrained and require further input, but the lifeguard problem does not.

*STEP* Do these three things: 1) Click in the Set Objective input field and select cell T2 by clicking on it. 2) Click the Min radio button. 3) Click the By Changing Variable Cells input field and select cell Q2. Click Solve. Click OK after Solver displays a dialog box that says it has found the answer.

Excel displays the best prediction, the one that minimizes the SSR, in cell Q2. Look at your scatter plot to see that this value really is at the very bottom of the U-shaped function.

But do you notice something else? Compare cell Q2 to the median and average cells you created earlier. That is amazing—the average is the best prediction!

It is true that given any list of numbers, the average is the value that minimizes the sum of squared residuals. Later, we will see how regression does the exact same thing.

## 1.5 Takeaways

IPUMS-CPS has the ASEC survey, which is the March CPS. It includes many questions on income and work. We used it to download data on income and education.

Income is not normally distributed. Instead of bell-shaped, it has a long right tail. Most workers earned around \$60,000 in income in 2023 (they were reporting last year's income in the 2024 survey). This is the median income.

The average was much higher than the median income because a few people made a lot more. The highest income in the sample was \$1,399,999. In the population, there are even higher incomes, with a few people making of tens and hundreds of millions of dollars in income. These high values increase the average, but the median stays unaffected.

While we often give the median as a measure of central tendency for distributions with long tails, it is the average that is the best prediction in the sense of minimizing the sum of squared residuals.

A residual is the difference between an actual and predicted value. The usual convention is to put a hat on predicted values, like this $\hat{Y}$. A residual is $Y - \hat{Y}$. A squared residual is then $(Y - \hat{Y})^2$ and the SSR is $\sum(Y - \hat{Y})^2$.

Most of this should be familiar (perhaps only vaguely so) if you took a statistics class. The histogram, average, and median are foundational concepts. Your stats professor might not have told you that the average minimizes the sum of squared residuals, but this is worth remembering.

# 2. The Conditional Mean Function

We continue working with IPUMS-CPS, which contains the ASEC (March CPS) survey. We are heading toward regression analysis, but we proceed slowly because we want to convey a crucial idea: *regression is closely related to the average.*

We will use Excel's powerful PivotTable tool. Microsoft trademarked this one-word term in 1994, but it was canceled in 2020—Microsoft continues to use the one-word version). We will also utilize VLOOKUP and SERIES functions.

Our key concept is the *conditional mean function* (CMF), which we can refer to more simply as a *graph of averages.* The CMF is one step away from regression and will help us understand how to interpret regression coefficients.

Recall that our research question is a true classic: *What is the rate of return on education?* Sure, most people believe that more education leads to higher income, but does it really? And, if so, by how much? Is it a simple linear relationship so another year of schooling raises income by a constant dollar amount or is something else going on?

We have explored INCWAGE, our dependent (Y) variable, and we have EDUC, our independent (X) variable, in our data set so you would think we could just dive in. We cannot, however, use EDUC in its raw form. The numbers represent categories so we cannot use them without doing some extra work. We have to recode this variable.

## 2.1 Constructing EducYears

*STEP* Open your CPS Excel file and examine the EDUC variable.

The numbers are clearly not years of schooling. No one goes to school for 81 years! Although numerical, these numbers represent answers to survey questions. But what exactly do these numbers mean? We have to return to the IPUMS-CPS website and look at the *codebook.* It has an explanation of each variable and a mapping of numbers to categories.

*STEP* Login at cps.ipums.org and click *MyData* (next to *Select Data*).

IPUMS provides a history of your data extracts, sorted by date. After a while, it removes the extract, but you can easily resubmit it if needed.

*STEP* Click the *Basic* link (under *Codebook*) and scroll down to see the codes for EDUC.

We will copy and paste code information into a new sheet in our workbook (Excel's *Text to Columns* is a nifty feature that will help us). Then we will add a column in the *rawdata* sheet which uses Excel's VLOOKUP database function to give us the number of years of schooling for a given level of EDUC.

The function has four arguments (or inputs): VLOOKUP(lookup value, table array, column number, range lookup). The fourth argument is optional and we will not use it. The function finds the lookup value in the table array and then reports value from the column number. This makes more sense (as usual) when you actually do it.

*STEP* Begin by adding a sheet to your Excel workbook and rename it *EducRecode*.

Next, we need to get the codes for the EDUC variable and put them in that sheet. This will be the table that we will use with the VLOOKUP function to construct the EducYears variable.

*STEP* Go to the IPUMS webpage with the Basic codebook and scroll to the EDUC variable, if needed. Select and copy from EDUC down to Missing/Unknown. Return to Excel and paste in cell A1 of the *EducRecode* sheet.

It looks good, but in fact, the information is all in column A. We need to separate the code from its description. This is a common issue when pasting data in Excel and separating cells with *Text to Columns* is worth remembering.

*STEP* Select column A (by clicking the letter A) and then click *Text to Columns* in the *Data* tab (in the *Data Tools* group).

This brings up a text-to-columns wizard dialog box. We use it to separate the code (in column A) and the descriptive text (which goes in column B).

*STEP* There are three steps to complete:

1. Select Fixed width (if needed) and click *Next*.
2. Excel may show a vertical line in position 16, right where the descriptive text begins, but if not, add one. Remove any other vertical lines. Click Next.
3. Click Finish.

Excel has split the combined information as we wished, but we have to add a new variable in column C that reports the number of years of schooling for each category.

*STEP* In cell C1, enter the label "EducYears" and in cell C2, enter 999 (the code for missing value). Figure 2.1 shows how to recode the other EDUC values into years of schooling (EducYears). We group those with less than ten years of education at ten years.

| EDUC | EducYears |
|---|---|
| 0,1, 999 | 999 |
| 2 to 50 | 10 |
| 60, 71, 72 | 11 |
| 70, 73 | 12 |
| 80 | 13 |
| 81, 90, 91, 92 | 14 |
| 100 | 15 |
| 110 to 122 | 16 |
| 123 | 18 |
| 124 | 20 |
| 125 | 20 |

Figure 2.1: Mapping EDUC codes to years of schooling.

With our table completed, we are now ready to use VLOOKUP (see the appendix to this chapter to try Excel's XLOOKUP function). Our strategy is to look up an EDUC code value in column A of our table and return the corresponding value in column C.

*STEP* Return to the *rawdata* sheet and enter "EducYears" as a label in cell AA1. Assuming your EDUC variable is in column M, in cell AA2 enter the formula: *=VLOOKUP(M2,EducRecode!$A$1:$C$37,3)* . Fill down. The dollar signs ensure that Excel uses the correct cells, always referring to range A1:C37, as you fill down.

Excel now displays in column AA the number of years of schooling that correspond to each observation's EDUC code value. This is exactly what we want.

This is a good time to document what you have done in your *doc* sheet and save the workbook.

## 2.2 Connecting INCWAGE and EducYears

We are now ready to do some data analysis. We begin with a histogram of our education variable.

*STEP* Select column AA, click the *Insert* tab in the Ribbon, and select *Histogram* from the *Charts* group. Double-click the $x$ axis and make the *Bin Width* 0.5.

The chart would need to be cleaned up if we wanted to use it in a paper or slide (it needs a title and axes labels), but it conveys several facts even in this primitive state:

1. There are relatively few observations with less than 12 years of schooling.
2. College graduates (16 years) are the most common observation, followed by a near tie between high school (12 years) and some college (14 years).
3. There are no people in this sample with 13, 15, 17, or 19 years of schooling.

But we are not interested in EducYears alone, we want to know how education affects income. We need to put them on the same chart.

Before we begin, we need to be aware of a few charting best practices. First, as with codes from a survey, a chart needs to be processed after Excel creates it. Second, Excel often makes default decisions that must be corrected and adjusted. Finally, try to be a minimalist when it comes to your graphs—avoid chartjunk, repeated text or numbers, and other distractions.

Creating a chart in Excel always requires three steps:

1. Select data (use the ctrl key when needed for non-contiguous data)

2. Click the chart type desired (usually Scatter and almost never Line or smoothed chart types)

3. Clean up (title, axes, and legend, if needed)

For a quick, 3-minute video of best-practice charting in Excel, check out vimeo.com/econexcel/how-to-chart-in-excel.

*STEP* Copy the INCWAGE column and paste it in column AB. Put your chart skills to work by making a chart of INCWAGE as a function of EducYears. Be sure to title the chart ("Income as a function of Education" is a good title) and label the axes.

What does your chart show? The answer is pretty much nothing. There does not seem to be much going on, just a bunch of points in a weird stacked pattern.

The chart is a series of vertical columns because EducYears has only a few integer values. It does not look like it, but the chart has tens of thousands of observations—many of them are repeated so there are many dots on top of each other.

We can improve the visualization and reveal a pattern if we compress the data. We can do this by finding and graphing the *average* INCWAGE for each value of EducYears.

*STEP* Select any cell in column AA or AB, then go to the *Insert* tab in the *Ribbon* and click *PivotTable* (usually in the top left). Confirm that the cell range is for the two columns of data and click OK.

Excel inserts a new sheet in your workbook with a blank PivotTable canvas.

*STEP* Click and drag EducYears into the *ROWS* area.

The PivotTable updates with our years of schooling variable as the left column of the table.

*STEP* Click and drag INCWAGE into the $\sum$ *Values* area. Click on the *Sum of* and select *Value Field Settings.* Choose the *Average* and click OK. Format the income cells in the table as $ with no decimal points.

This is an important table. It is displaying the average income given years of education. Since it is an average given a specific value for another variable, it is called a *conditional mean.* The table provides evidence that yes, in fact, average income rises as education increases. This is the beginning of an answer to our research question!

We want to make a chart of the data in the PivotTable, but we also want to compare it to the scatter plot of all of the data.

*STEP* Copy the Scatter chart you made in the *rawdata* sheet, return to the *PivotTable* sheet, and paste it.

Next to it, we want to put a chart of ***average*** INCWAGE as a function of EducYears. Excel, however, does not allow a scatter chart to be made directly from a PivotTable, but we can edit the SERIES function in a previously made chart as a workaround. Before attempting this advanced approach to charting, it is worth seeing it in action.

*STEP* Watch this short, 5-minute video at vimeo.com/econexcel/using-series-formula.

Now that you know how to directly edit the SERIES formula, we can proceed.

*STEP* Paste the Scatter chart from the *rawdata* sheet again (it should still be in your clipboard so a simple *ctrl-v* should do it). Directly edit the SERIES formula, replacing the $x$ and $y$ arguments with the cell addresses from the conditional mean table on the PivotTable sheet.

The chart you just made is showing what is technically called a *conditional mean function,* but a simpler name for it is a *graph of averages.* It shows the average income conditional on the years of schooling. Given a value of education, the CMF reports the average income just for that value.

The CMF clearly shows that there is a relationship between educational attainment and income. As schooling rises, so does average income.

In addition, the relationship is not linear. We can clearly see that income does not increase by a constant dollar amount for each additional year of education. For low levels of education (below a high school degree), another year of education does not seem to increase average income at all. At higher levels of education, income seems to takeoff. The graph of averages hints at a more complicated, non-linear relationship between education and income. We will explore this more carefully in future work.

The two charts are obviously closely related. The graph of averages is compressing each education level (on the $x$ axis) into a single point, the average for that value of years of schooling. To emphasize this, we can add the CMF, the average income values from the PivotTable, to the other chart. To learn a little more Excel, we will do this a new way.

*STEP* Select the *Income as a function of Education* chart, then click *Select Data* (in the *Design* tab). Click *Add* to get the *Edit Series* dialog box and fill in the three inputs. The Series Name is the cell that has the "Average of INCWAGE" in the data that was used to create the CMF chart. The *Series X* and *Y values* are the education years and average INCWAGE cell ranges. Click OK. Change the CMF series to a solid line by right-clicking it and using *Format Data Series*.

Your result should look like Figure 2.2.

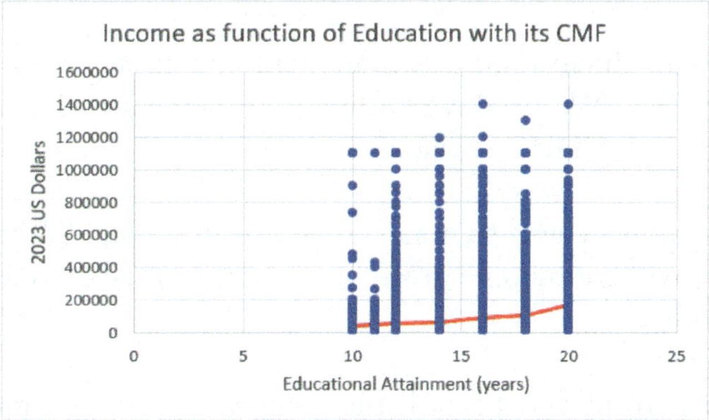


Figure 2.2: Scatter plot with its CMF highlighted.
Source: Data from 2024 IPUMS CPS.

Excel superimposes the graph of averages over the individual data. The average is much closer to the $x$ axis than in the standalone CMF chart to the right because each vertical strip has many more dots at the bottom than the top—remember, income has an extremely long tail.

Figure 2.2 makes it clear that the CMF, or graph of averages, is a compression of the data. Just like the average, the CMF is a summary of the data. It can reveal patterns in the data, but it also ignores variation in individual observations.

## 2.3 Takeaways

We found evidence, using IPUMS-CPS data, that people with more schooling earn higher incomes in the United States, on average. This does not mean, of course, that education caused higher income. We have to do a lot more work and control for other factors before we can come anywhere close to making such a claim.

A scatter plot of all of the income and education data does not show anything because there are so many dots (and many on top of each other) that it is impossible to see what is going on.

The CMF, a plot of average income for each number of years of schooling, does reveal the relationship: more education leads to more income, on average. Furthermore, the relationship is a curve. So more education does not lead to a constant additive increase, the increases get bigger and bigger. This is important.

Before we can use educational attainment in our analysis, we need to recode the EDUC variable. We used VLOOKUP to compute the number of years of schooling for each EDUC category.

PivotTables are an easy way to make crosstabs in Excel. We use a PivotTable to compute the average income given the number of years of education. This is a CMF in table form.

We use the data from the PivotTable to make a chart which displays the CMF as a set of points. If you connect the points, it will usually look like a squiggly line or curve.

*EXCEL TIP* Excel charts are actually formulas and the SERIES function has the following arguments:

=SERIES(legend text, x axis range, y axis range, series #).

SERIES formulas can be copied and pasted just like other objects and this gives us a powerful way to make and edit charts.

The CMF is a compression of the data. It is a conditional average telling you the average for a given value of x. This is important to remember because regression is going to do something similar.

## Appendix

VLOOKUP is an older Excel function that is backwards compatible with versions of Excel before 2021. This is why it is still used.

More modern versions of Excel have a new function, XLOOKUP, which is better and works in a similar way. The syntax is =XLOOKUP(lookup value, lookup range, return range), ignoring several optional arguments.

To try it, use this formula in your CPS workbook:
=XLOOKUP(M2,EducRecode!A:A,EducRecode!C:C).

VLOOKUP requires the entire table array with the values you want to lookup in the leftmost column and the column number from which you want the output value. Leftmost is clunky, but worse is the column number requirement. This is not good for flexibility. If you insert a column in the table you will break the VLOOKUP function.

XLOOKUP is definitely better. It is faster, more reliable, more flexible, and easier. Its only drawback is that you need newer versions of Excel. If your workbook might be used by someone with older Excel, consider VLOOKUP.

Believe it or not, there is an even older function, LOOKUP, but you should never use it because it has a confusing vector versus array structure.

Excel was first released on Apple's Macintosh in 1985 and a Windows version followed two years later. Since that time, it has evolved and grown, but many old functions and components remain so that users do not have to constantly update their files.

Translating information from one variable to another is an exceedingly common task. It is good to know about VLOOKUP and XLOOKUP and how they work.

# 3. The Mechanics of Regression

Regression is the most common data analysis in economics. Loosely speaking, it involves fitting a line to a cloud of points. Our focus here is on understanding exactly how the line is fitted and the coefficients are found.

The actual process of fitting the line is different from the interpretation of regression. Understanding how regression coefficients are computed will help us master what they mean and how they are used. Our work will highlight LINEST, an Excel array function, and Solver, Excel's optimization add-in.

## 3.1 Running a Regression in Excel

We begin by entering some data. This gives us a concrete problem to work on.

*STEP* In cells A1 and B1 of a blank Excel workbook, enter the labels *X* and *Y*, respectively. In cell A2, enter a 1, followed by a 2, in cell A3. Select cells A2 and A3, then fill down to A11 so you have a series from 1 to 10. For the Y data, enter the formula *=10 + 5*A2 + RAND()*20* in cell B2, then fill it down.

Column B's values will change every time the spreadsheet is recalculated. This is good for simulation, but bad for this application. We need to deaden column B so that we can use Solver.

*STEP* Copy cells B2:B11, click the down arrow in the Paste button (on the *Home* tab in the Ribbon), and select Paste Values. Click on a cell in the range B2:B11 and confirm that you have replaced the formula with a number.

Excel offers several ways to run a regression. Perhaps the easiest is to fit a line in a chart.

*STEP* Create a Scatter chart and then fit a trendline by right-clicking one of the dots in the chart and selecting *Add Trendline*. You do not need axes labels or a legend, but the title should be *Y on X* which is short for "regress Y on X." This is how economists speak—it means that Y depends on and is a function of X.

Excel displays options for the trendline you requested.

*STEP* Confirm Linear (the default) is selected and check *Display Equation on chart.*

Congratulations! You just ran your first regression in Excel. The equation of the line, $y = b_0 + b_1x$, has an intercept ($b_0$) and a slope ($b_1$). We need to understand exactly where these coefficients come from, but before we work on that, we will run the regression a second way, with the LINEST array function.

While LINEST is easy once you get used to it, at first, it can be frustrating and confusing. It is a primitive function in the sense that there are no bells and whistles or friendly user interface. But it is powerful and well worth the effort to master it.

In Excel, array formulas span more than one cell, and the results are displayed in more than one cell. Because LINEST is a formula, Excel will update the results whenever the sheet recalculates, which is perfect when we have live cells with RAND() and run simulations.

The LINEST function has four arguments, each separated by a comma.

1. Y values cell range
2. X values cell range
3. TRUE (or 1) to include an intercept in the regression or FALSE (or 0) if you want the line to go through the origin
4. TRUE (or 1) to report other regression results besides the slope and intercept or FALSE (or 0) to suppress additional regression output

The formula *=LINEST(1, 2, 3, 4)* is entered into a previously selected cell range by a special keystroke combination, ctrl-shift-enter.

To use LINEST, you must select the appropriate cell range in which to place the output. The number of columns selected should equal the number of coefficients to be estimated. The number of rows to select depends on the output you want. If you just want the regression equation coefficients, the selected range should contain only one row. If you would like the full set of regression results, then the range must be five rows tall.

*STEP* Below the data, select a range of cells that is 5 rows by 2 columns. Five rows because we want all the output (we will use a 1 for the fourth argument) and two columns because we have an intercept and a slope (we will use a 1 for the third argument).

With the appropriate cell range selected, you are ready to type in the LINEST function and input the necessary information.

STEP With the cells selected, go to the formula bar and enter the formula: *=LINEST(B2:B11,A2:A11,1,1)*, then hold down the ctrl and shift keys and press enter.

Excel displays ten numbers in the 5 by 2 range. Once the LINEST array formula is entered, all of the cells in the range have the same formula within a pair of curly brackets, .

Because LINEST is an array formula that spans several cells, you cannot edit the formula by clicking in a single cell and hitting the enter key. If you want to change the array formula, you must press ctrl-shift-enter.

You can always hit the aptly named escape key, esc, to get out of a cell with an array formula in it. **Remember, if you get stuck, press the *esc* (escape) key to get out.**

But what exactly do the numbers mean? LINEST is so raw and primitive that it has no descriptive labels or explanation. The user has to know the map of the output, shown in Figure 3.1.

| $Y = b_0 + b_1 X$ | |
|---|---|
| $b_1$ | $b_0$ |
| SE($b_1$) | SE($b_0$) |
| $R^2$ | RMSE |
| $F$ | df |
| RegSS | SSR |

| $Y = b_0 + b_1 X_1 + b_2 X_2 + b_3 X_3$ | | | |
|---|---|---|---|
| $b_3$ | $b_2$ | $b_1$ | $b_0$ |
| SE($b_3$) | SE($b_2$) | SE($b_1$) | SE($b_0$) |
| $R^2$ | RMSE | #N/A | #N/A |
| $F$ | df | #N/A | #N/A |
| RegSS | SSR | #N/A | #N/A |

Figure 3.1: LINEST output.

The left panel in Figure 3.1 is for a bivariate regression (two variables, like we have in this example) and the right panel is for a multivariate (also called a multiple) regression. Right now, we care only about the top and bottom rows in the left panel, but it is no exaggeration to say that mastering all ten of the cells is the goal of most econometrics courses.

*STEP* Compare the top row of your LINEST output to the equation of the line on your chart to confirm that they are identical.

One quirk of LINEST to remember is that it outputs the coefficients backwards, so the intercept is in the top right corner and the slope coefficients are displayed in reverse order (see Figure 3.1).

But where exactly do the coefficients come from? We turn to a third way to run a regression in Excel to help us answer this fundamental question. We will use Excel's Solver to run a regression, getting the same coefficients, of course, but revealing exactly how the line is fit along the way.

*STEP* Remove the trendline from the chart. Right-click and select *Delete*.

Our strategy will be to input intercept and slope coefficients and create a formula for a line on the spreadsheet, then add it to the chart.

*STEP* In cells D1 and D2, enter the labels *b0* and *b1* and name cells C1 and C2, *b0_* and *b1_*. (The underscore character is needed to distinguish the name from cell B1.) In cell E1, enter the label *Predicted Y* (also known as Y-hat, $\hat{Y}$) and enter the formula *= b0_ + b1_*A2* in cell E2, then fill it down.

Excel shows zeroes because of the blank cells in C1 and C2. This is one of the most dangerous things a spreadsheet does. It should display a missing value or other warning to the user when a cell is an input and is blank. Even more troubling is that a spreadsheet does not treat blanks as zeroes all the time—in some functions (like the average), they are not treated as zero, but ignored.

This cavalier treatment is an example of why some data scientists shout, "Friends don't let friends use Excel!" Spreadsheets are, however, ubiquitous and a dominant data tool in the workplace. A better response is to learn their weaknesses to avoid making devastating errors.

*STEP* Display the data in column E in the chart by copying the SERIES formula in the chart, pasting, and editing it. With the series just added selected on the chart, click *Change Chart Type* (in the *Design* tab) and make the series a scatter with straight lines and no markers.

To get the line off the ground, we change the intercept. Pick a value for the intercept roughly around the average Y value in your data.

*STEP* Enter this number (use 50 if you are unsure) in cell C1.

Excel lifts the line up in the chart, showing a horizontal line at the value of the intercept you entered. This makes sense, since the slope is still zero (the interpretation of a blank cell), the line is flat.

This is obviously a poorly fitted line. Can you do better by playing around with the intercept and slope to best fit the line?

*STEP* Enter values in cells C1 and C2 until you have what you think looks like the best fitted line.

This is not, obviously, how regression lines are fitted to data. Instead, they are the product of solving an optimization problem: choose coefficients to minimize the sum of squared residuals (SSR).

We have seen this before, remember, when we showed that the average is the best predictor in the sense that it minimizes the SSR. We go slowly, however, to be sure there is no misunderstanding.

A *residual* is simply the vertical distance from an observation to the fitted line. The residual is the actual value of Y minus its predicted value.

*STEP* In cell F1, enter the label *Residual* and in cell F2, enter the formula = *B2 - E2*.

The number displayed in cell F2 is the residual for that observation. The bigger the residual, the farther away the actual value is from the fitted line.

*STEP* Fill down the formula in cell F2 to F11.

Depending on how well you fit the line, you should see some positive and some negative residuals. The positive ones are where the actual value is above and the negative ones are where Y is below the fitted line. If $Y = \hat{Y}$, then the residual is zero and the point is exactly on the regression line.

Next, we square the residuals. This makes them all non-negative. It also exacts a harsher penalty when an observation is far from the fitted line—after all, squaring is an operation that makes all positive numbers get larger, but multiplying a big number by itself has more an effect than squaring a smaller number.

STEP In cell G1, enter the label, *Squared Residual* and in cell G2 enter the formula *=F2^2*. Fill it down.

Finally, we need to add up the squared residuals.

STEP In cell G12, enter the formula *=SUM(G2:G11)*. Next to it, enter the label *SSR*.

Cell G12 computes the sum of squared residuals for the coefficients in cells C1 and C2. If either of those cells changes, the SSR changes. We now have a metric with which to judge how well the line fits the data. Our goal is to minimize the sum of squared residuals.

STEP Play around with cells C1 and C2 to see if you can lower cell G12. You are searching for the best intercept and slope, where "best" is defined as the pair that produces the lowest possible sum of squared residuals.

Manually hunting and pecking in cells C1 and C2 might help explain how the coefficients determine the line and value of the SSR, but it is a poor way to find the actual optimal solution. A much better way is via Excel's Solver.

Solver is a numerical optimization algorithm that is part of Excel. It lives in the *Data* tab. If it is not there, you must install it with the Add-ins Manager (File → Options → Add-ins → G0 or keyboard shortcut *alt*, *t*, *i*.).

STEP Click the *Data* tab and click the *Solver* button.

Excel displays the Solver Parameters dialog box. This is where you give Solver the information needed to process the problem. Solver will try values in the *Changing Variable Cells* that most improves the *Target* cell (or objective function), subject to given constraints.

STEP Make the *Target* cell $G$12 (it is easiest to click on cell G12), click *Min*, and make the Changing Variable Cells $C$1:$C$2. Click *Solve*.

Solver displays a dialog box that says it has "converged to the current solution." You can see that cells C1 and C2 have been changed.

STEP Click OK and click on cells C1 and C2 to confirm they are machine-generated, with many decimal places.

Perhaps you noticed that LINEST and Solver give the same coefficients, but, if not, compare the intercept and slope produced by Solver to that of LINEST. They are practically speaking the same, but not exactly the same.

STEP Widen columns A, B, and C and show ten or so decimal places. Compare Solver and LINEST to confirm the two are quite close, but not exactly the same.

Solver continues trying values until it hits its convergence criterion. Most of the digits displayed in cells C1 and C2 are *false precision.* It seems like an incredibly precise answer, but it is not.

STEP Another important number to compare is the SSR in cell G12 to bottom-right corner cell of LINEST (where it displays the SSR).

Like the coefficients, the two SSRs are almost exactly the same. LINEST and Solver are giving us the same result, practically speaking. Unlike Solver, LINEST is using an analytical solution produced by solving the optimization problem with calculus and algebra. This is the problem:

$$\min_{b_0, b_1} SSR = \sum_{i=1}^{n} (y_i - \hat{y}_i)^2 = \sum_{i=1}^{n} \left(y_i - (b_0 + b_1 x_i)\right)^2$$

We care more about how the regression line is determined than the mathematics of how to actually find it. You need to know LINEST is giving us the best fit based on the solution to an optimization problem that is obtained by an analytical expression. LINEST is not using Solver.

A 3D visualization of the optimization problem makes clear what regression is actually doing. We will use Excel's Data Table tool (once again) to create data, then make a chart. The Data Table tool is powerful, but can be confusing. As you perform the steps, ponder what is being constructed.

*STEP* In cell D18. enter the formula =*G12*. Create a series from 0 to 50 by 10 in cells E18:J18. Create a series from 0 to 10 by 1 in cells D19:D29.

Excel will input these row and column values and output the SSR for these pairs in the cells in the table.

*STEP* Select cell range D18:J29 and click *Data Table* in *What-If Analysis* in the *Data* tab. In the row input field, click or enter C1, and in the column input field, click or enter C2. Click OK.

Excel populates the inside of the table with SSR of each corresponding intercept-slope pair.

*STEP* Enter a 10 in cell C1 and a 1 in cell C2 to see that cell G12 equals the value for that pair in the table.

To make a 3D chart, we need a blank cell in the top left corner of the data. Then we can apply the three steps to make a chart: (1) select data, (2) choose a chart type, and (3) clean up.

*STEP* Click on cell D18 and drag it up to cell D17. Select cell range D18:J29 and click on *Recommended Charts* in the *Insert* tab.

We want a Surface type chart.

*STEP* If the Surface type is not listed, click on the *All Charts* tab and select *Surface* and *3-D Surface*. Click OK.

Excel places a chart on the sheet, but we are not done. To make a chart in Excel, we always want to complete the third step: adjust what Excel has created.

*STEP* If Excel has added a legend, remove it. Add *min SSR* as the title and label the axes as *slope (b1)*, *intercept (b0)*, and *SSR*. Double-click the SSR (vertical) axis and set the *Major Units* in the *Axis Options* to 1000.

This last adjustment has a dramatic effect on the chart. As shown in Figure 3.2, it is clearly a valley with a distinct minimum. The intercept-slope pair that best fits the line is giving us this minimum point.

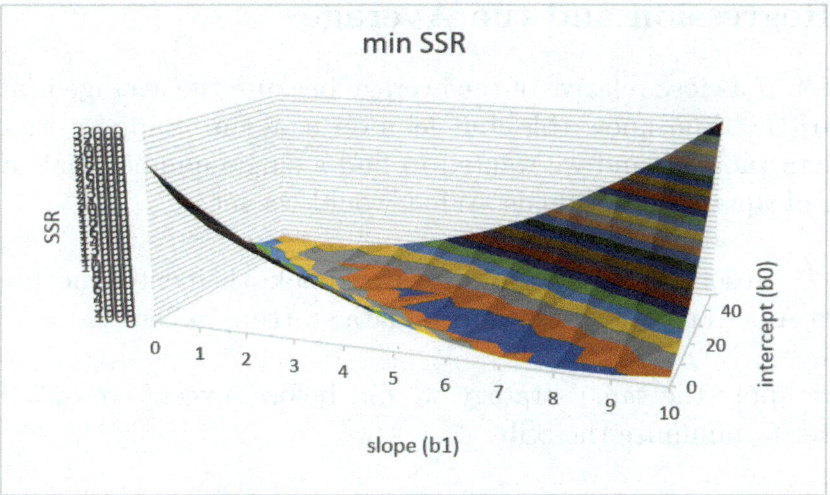


Figure 3.2: Minimizing the SSR.

The shape of the valley (or bowl) in Figure 3.2 and its height at the optimal solution (the minimum SSR) depends on the data. If all of the observations lie on the regression line, the SSR is zero and we have a perfect fit. The farther away the dots, the higher the minimum SSR of the regression line and the worse the fit.

## 3.2 A Big Misunderstanding

After learning how the regression line is fitted, many students jump to an erroneous conclusion: "Regression is used because it minimizes the SSR." This is a big mistake.

Regression is a recipe and it is true that it is based on minimizing the SSR, but this is not why it is used. There are, in fact, many other possible recipes (for example, minimize the sum of the absolute value of the residuals).

The reason why we choose to minimize the SSR is too complicated to explain right now, but it is fundamental to econometrics and will be explained in future work. It involves the names of two of the greatest statisticians of all time, Gauss and Markov. The Gauss-Markov Theorem, which says that OLS is BLUE, is the reason why we choose to minimize the SSR instead of doing something else. Explaining what "OLS is BLUE" means takes a lot of work, but it is fundamental and we will take the time to explain it in detail when we work on regression for inference.

## 3.3 Regression and the Average

Regression is a close relative of the average because the average is also a min SSR statistic. You know this, but let's do it again. Suppose we only had the $Y$ data (no $Xs$) and we wanted to find a single number that minimized the sum of squared of residuals. What would we get?

*STEP* Insert a new sheet in your workbook, then enter the label *Actual Y* in cell A1. Copy and paste your Y data starting in cell A2.

Next, we apply the same strategy we did before, except we only have one parameter to minimize the SSR.

*STEP* In cell C1, enter the label *coeff.* In cell D1, enter the label *Predicted Y* and enter the formula = *$B$1* in cell D2, then fill it down.

Our job is to find the "best fit" value of coeff. We use minimize SSR as an objective measure of best fit.

*STEP* Create residuals in column E, then square them in column F and add them up in cell F12.

You can hunt-and-peck in cell B1 to see how low you can drive cell F12, but this is obviously a job for Solver.

*STEP* Run Solver to find the value of cell B1 that minimizes the SSR in cell F12.

Just as we did before with a Data Table, we can make a chart of what we are doing. But we can do it a different way.

*STEP* Select an empty cell. Insert a *Scatter* chart. Select the blank chart and enter this SERIES formula:
=SERIES(,{0,0,0,0,0,0,0,0,0,0},Sheet2!$A$2:$A$11,1).

Notice we gave the chart data in a new way, entering ten zeroes in curly brackets. Excel produces a chart with the ten data points in a vertical line at $X = 0$. Cell B1 gives the number that minimizes the SSR and gives the "best fit" single point. What is this point?

*STEP* In an empty cell, enter the formula: =AVERAGE(A2:A11).

You are not surprised because you knew this already. The average Y and the value that minimizes the SSR are the exact same number! There is no doubt about it: regression and the average are intimately related. In fact, regression is just an extension of the average.

## 3.4 Takeaways

Perhaps you have heard of regression referred to as *least squares*? This is how regression was first described in the late 1700s and early 1800s when it was first discovered. There was a bitter fight over who did it first. "It is argued (though not conclusively) that Gauss probably possessed the method well before Legendre, but that he was unsuccessful in communicating it to his contemporaries" (Stigler, 1981, p. 465).

"The method" that Gauss had was that you find the line of best fit by minimizing the sum of squared residuals. Today, we let software almost instantly and effortlessly produce regression coefficients, but it was a tremendous accomplishment to figure out the correct way to process the data (sum the squared residuals) and solve the optimization problem (least squares).

We ignore the mathematics behind the analytical solution. Instead, we focus on understanding that the coefficients are determined by a procedure that minimizes the sum of squared residuals. That is a key takeaway.

Modern versions of Excel allow you to use just the enter key to enter an array formula like LINEST. The first time you do this, Excel pops a message saying spillover cells are being displayed. This mimics the behavior of Google sheets (who did this first). It is best practice to be aware of what an array formula is and how it works. Learning ctrl-shift-enter is a good thing.

## References

Stigler, S. (1981) "Gauss and the Invention of Least Squares," *The Annals of Statistics*, pp. 465–474, www.jstor.org/stable/2240811.

# 4. Interpreting Regression

You run a regression and you get results. Excel's LINEST array function spits out 10 pieces of information, but there are many, many more diagnostics and statistics. In this section, we will work on understanding the most fundamental of regression's results, the coefficients. We will also explain why it is called regression, which you have to admit is a pretty odd name.

## 4.1 The Equation of the Regression Line

*STEP* Return to your *rawdata* sheet in your IPUMS CPS Excel workbook, right-click a point on your chart of INCWAGE as a function of EducYears and fit a (linear) trendline (display its equation on the chart).

Excel displays the equation of the fitted (regression) line in the common math format of $y = mx + b$. The coefficients in the equation are the intercept (which is negative in our case) and the slope.

For the 2024 ASEC, the regression equation is

$$PredictedINCWAGE = -79814 + 11073EducYears$$

*STEP* Use LINEST to replicate the coefficients displayed on the chart.

From previous work, we know how the regression line is actually fitted—the coefficients are chosen to minimize the sum of squared residuals. But what can we do with this equation? Most importantly, we can *predict income given education.*

The CMF makes a similar prediction. It tells us the average INCWAGE given a level of schooling. Each conditional average is isolated, computed from only the observations that have a particular education level. For high school grads, the CMF says average income is \$57,297.

The regression equation, like the CMF, can be used to tell us how much high school grads make. We simply plug in the number 12 for EducYears and compute Predicted INCWAGE:

$$PredictedINCWAGE[EducYears = 12] = -79814 + 11073[12]$$
$$PredictedINCWAGE[EducYears = 12] = 53060$$

The regression's predicted income for high school grads is close, but a few thousand dollars less than average income for high school grads. Unlike the CMF, which uses just the people with a high school degree, the regression equation uses *all* of the data.

*STEP* Return to the PivotTable sheet and add a label, "Predicted INCWAGE", next to "Average of INCWAGE" in the PivotTable. Below it, add a formula that computes predicted income given years of schooling. Use the coefficients from LINEST to build your formula.

For each year of education, predicted income from regression is a little different, but pretty close to the CMF's conditional average, except for the last value. But the next step will show these two are intimately related.

*STEP* Add the Predicted INCWAGE values to the CMF chart (copying and editing the SERIES formula is an easy way to do this).

Figure 4.1 shows the result and points to a key concept: we can think of the regression line as *linearizing the squiggly graph of averages.* We postpone for future work the fact that a line is a poor description of the relationship of income and education since it is obviously curved.

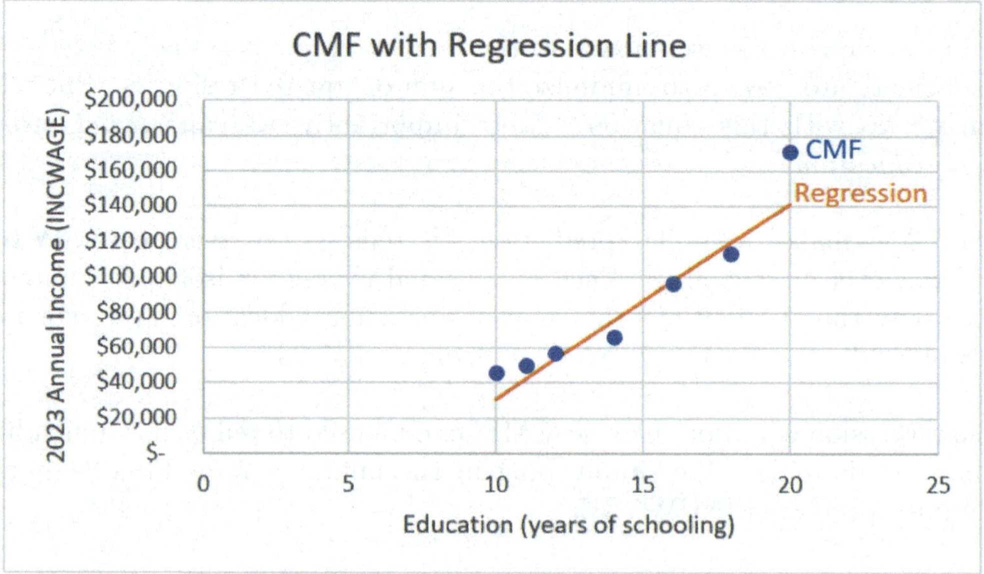


Figure 4.1: Interpreting regression as a conditional average.

Thus, regression is a second compression of the data. The first compression is the CMF, taking vertical strips and finding the average within the strip. Then, a further compression occurs when the dots of the CMF are linearized.

Note that this idea of a double compression is a way to understand what regression is doing, but it is not how the regression line is actually fitted. Regression is not computed via the CMF, but we can view the regression equation as a linearization of the CMF.

This interpretation ties regression directly to the conditional average. Recall our work showing that the average is the best predictor (in the sense of minimizing the sum of squared residuals) of a list of numbers. Regression is an extension of that. If you were asked to predict a Y value, you would use the regression line because it is the best linear predictor in the sense of minimizing the SSR.

## 4.2 Interpreting the Coefficients

Next, we turn to the meaning of the coefficients themselves. The intercept is usually needed only to place the line and often does not have an important interpretation. This is true in this application.

The intercept gives the predicted income of someone with zero years of education. In this case, that would be negative $79,814. That does not make much sense. The intercept is "far" from the data cloud and the regression equation is making a terrible prediction for such a situation. We would not use regression to predict income for zero education.

The slope coefficient is almost always the star of the show. What does it tell us? A simple computation will immediately reveal its meaning.

*STEP* Add another column next to Predicted INCWAGE in your PivotTable with the label "DINCWAGE." Select the letter D in the formula bar and change the font to Symbol. This makes the character a Greek letter D, $\Delta$ (delta), which means "change." Two cells below, enter a formula that computes the difference in Predicted INCWAGE from 10 to 11 years of schooling.

The change in Predicted INCWAGE from 10 to 11 years of education is $11,073 which is exactly the slope coefficient. Does this work for other values?

*STEP* Fill your formula down.

The value doubles for those years that skip a value because the years of education also double. The slope coefficient tells you *the rate of change of Predicted INCWAGE for a given change in EducYears.*

This "rate of change" language may trigger thoughts of the derivative.

$$PredictedINCWAGE = b_0 + b_1 EducYears$$
$$\frac{dPredictedINCWAGE}{dEducYears} = b_1$$

Thus, the slope coefficient is a measure of the change in predicted income as education changes. Note that the units of the slope are dollars per year of schooling. When we multiply by EducYears, which has units of years, we get dollars. The intercept's units are always the same as the dependent variable's unit (in this case, dollars).

Usually, most of the attention in a regression is focused on the slope coefficient because it tells us how much Predicted Y changes as X changes. In the 2024 ASEC data, an additional year of education was associated with $11,000 of extra income. That is a lot when you consider that you will probably work for 40 years or so. But when we compare a college versus a high school degree, the difference gets even bigger.

How much more does the regression predict a college graduate will earn than a high school grad?

There are two ways to answer this question and they mirror the two interpretations of regression we have discussed:

1. Plug in 12 and compute predicted INCWAGE; then repeat for 16 years of schooling. Difference the two outputs.

2. Multiply the slope by four—if $b_1$ tells us the rate of change for one year, $4b_1$ should tell us the change over four years.

*STEP* Do these two methods give the same answer?

Of course they do and we can use this fact as a check in future work. College is expensive and more is needed to measure the return on education, but making an additional $44,000 per year over a lifetime seems like a good reason to acquire a college degree. We will see, however, that this number is a poor estimate of the returns to education. It surely suffers from *omitted variable bias*—that sounds and is bad, but we must postpone explaining it because it is quite complicated.

We have shown that one can "read" the regression equation as "predicted Y given X" and the slope coefficient tells us "how much Y rises when X rises." The impact on Y can be computed by a calculation of the change in predicted Y from one point to another or via the slope.

## 4.3 What's in a name?

But why is it called regression? Least squares tells us the recipe being applied (minimize the sum of squared residuals). Where does the name regression come from and does it make any sense?

The British polymath (and eugenicist) Sir Francis Galton popularized the term. He published a paper titled "Regression towards mediocrity in hereditary stature" in 1886 and used the words *regression* and *reversion* to describe his key finding. Galton wanted to know if tall dads had sons as tall as them and the same for short people—are their sons as short as their fathers?

Karl Pearson, Galton's protege and biographer, extended Galton's work on height. A modified version of Pearson's data is shown in Figure 4.2 (from www.randomservices.org/random/data/Pearson.html). There are 1,078 heights of British fathers (of various ages and backgrounds) and their sons.

The father's average height is just under 68 inches, while the son's average height is just under 69 inches. The intersection of the solid horizontal and vertical lines in Figure 4.2 shows the point of averages, or center of the cloud of dots. The standard deviation (SD) tells us the dispersion and the correlation coefficient (r, invented by Pearson) measures the strength of association.

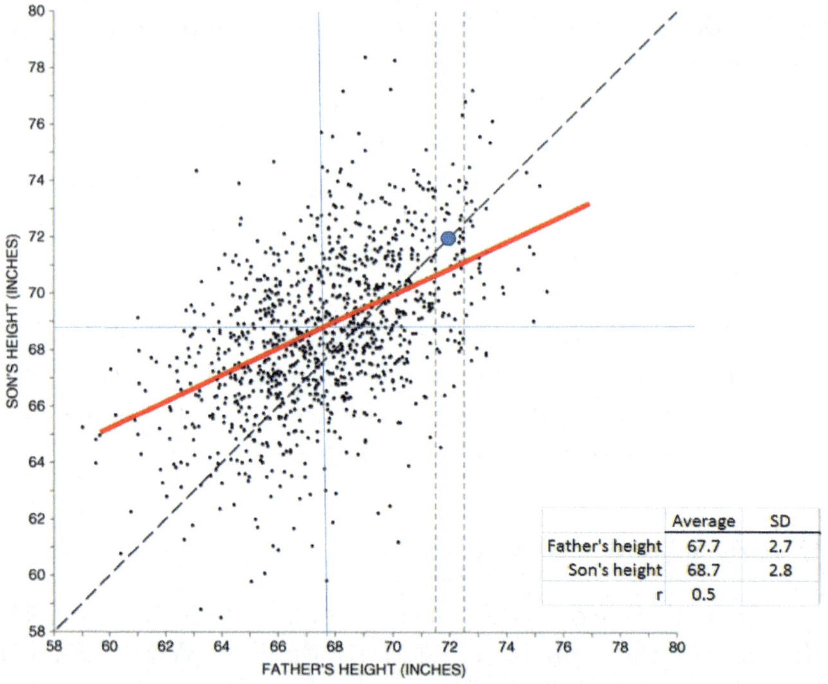


Figure 4.2: Father's and son's heights.

Notice that the point of averages is above the dashed 45 degree line where the heights are equal. The red regression line goes through the point of averages so it is always true that predicted Y given average X is average Y.

*STEP* Confirm this claim for the regression of INCWAGE on EducYears.

The vertical strip at 72 inches ($\pm 0.5$ inches) shows a blue dot to mark the place where son's height is also 72 inches. But is 72 inches the average height of sons whose fathers are 72 inches tall? No. The blue dot is too high up to represent the average height for these sons with taller than average dads.

The regression line at 72 inches would be a much better predictor of son's height. It is obviously lower than 72 inches. So, the son's of tall dads are taller than average (above the horizontal line), but they fall back to the pack (below the blue dot) and this is why the term regression stuck.

It works the same way for shorter than average fathers.

*STEP* Look at 64 inches on the $x$ axis in Figure 4.2 and go straight up. The average son's height for this vertical strip is not 64. That would be too low. It looks like it is around 67 (where a vertical line at 64 would intersect with the red regression line).

Again, this illustrates regression or reversion to the mean. The sons of shorter than average dads are not as short as their dads—they are shorter than average, but closer to the overall average height than their dads.

It is called regression analysis because of the fact that many applications contain this type of data cloud that is oriented so that average Y conditional on higher or lower than average X regresses (or reverts) back to mean Y.

The most common way to get this effect is when randomness is involved. This applies to height because there is a genetic lottery at work as genes are passed down. So really tall people are that way partly because they happened to get DNA sequences that express tallness, like rolling a 12 with two dice. Their kids will be tall, but not as tall as their really tall parents because the kids are unlikely to roll a 12 again. The fancy name for this effect is regression to the mean and it is how regression analysis got its name.

## 4.4 RMSE and $R^2$

Although not part of interpreting the coefficients, there are two basic diagnostics reported by regression packages, including LINEST.

The root mean squared error, RMSE, does for regression what the SD does for a list of numbers: it measures dispersion. The RMSE is poorly named, it should be called the root mean squared residual, not error.

To compute it exactly correctly is complicated (there are issues of degrees of freedom as in the difference between the sample and population SD), but the RMSE is fundamentally the SD of the residuals. Thus, it measures the dispersion of Y around the regression line.

The RMSE even follows the SD's rule of thumb that about 68% of the observations are within $\pm 1$ SD from the average and 95% are within $\pm 2$ SDs, when the data are normally distributed around the regression line.

For example, in Figure 4.2, the RMSE is about 2.4 inches (its units are those of the Y variable). If you drew lines with the same slope as the regression line, but shifted up and down by 2.4 inches, about 68% of the dots would fall in that band. About 95% of the observations would fall in a wider band that was 2 RMSEs above and below the regression line. Interpreting the RMSE as bands around the regression line is shown in Figure 4.3.

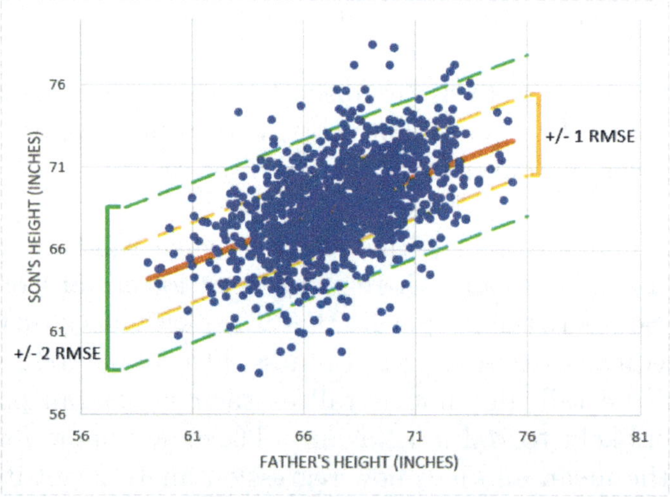


Figure 4.3: RMSE bands around the regression line.

*STEP* LINEST reports the RMSE in the 3,2 position. Return to the CPS data sheet to confirm that the RMSE is $81,703.

This means that there is a lot of dispersion around the regression line and tells us that income depends on much more than just education. That makes sense.

A second common regression diagnostic is $R^2$. LINEST reports it in the 3,1 position of its output. It tells us the percentage of variation in Y that is explained by the regression. It can vary from zero to one (or 100%).

*STEP* Confirm that the $R^2$ is 0.1 for the regression of INCWAGE on EducYears.

This is quite low. It agrees with the RMSE evidence that education alone is not explaining a lot of the variation in Y.

The $R^2$ statistic is known as a measure of fit. It is computed by subtracting $\frac{SSR}{TSS}$ from 1, where TSS is the total sum of squares. LINEST reports the regression sum of squared residuals in position 5,1.

*STEP* In a cell below LINEST, add the two cells in the bottom row of LINEST. This is TSS. Next, create a formula that computes $1 - \frac{SSR}{TSS}$. Confirm that this agrees with LINEST's 0.1 value.

The $R^2$ when we regress son's height on father's height in Figure 4.2 is 0.25. That is not great either—apparently, a lot of other things besides a father's height determines a son's height.

## 4.5 Takeaways

A regression equation can be interpreted as a linearization of the CMF. This directly ties regression to the average.

Just like the CMF, by plugging in a value for X, a regression equation provides a predicted value for Y.

The intercept is usually not our primary focus. It serves to place the line.

The slope is almost always the coefficient we care about the most. It tells us the rate of change in Y given a change in X. As such, it is a measure of the impact or effect of X on Y.

We often give the average as a representative number for a list and we give a slope coefficient as a measure of the relationship between X and Y.

Two other regression diagnostics are RMSE and $R^2$. Both are trying to indicate how well the regression is performing. We want the RMSE to be low and the $R^2$ close to one. There is, however, no single measure that tells us if a regression is good or bad. Such an evaluation requires judgment and more information.

The application of line fitting to father's and son's heights is how regression got its name. Son's height regresses to the mean. The source of the name is not so important, but it is nice to know.

Figure 4.4: Francis Galton (1822-1911) generated by ChatGPT.

Galton, Pearson, and others interested in genetics wanted to know how height was passed on. They found out through data analysis that tall dads had taller-than-average sons, but the sons were not as tall as their tall dads. Height provides a way to measure public health and economic development. It is partly heritable, but also affected by one's environment.

It is surprising how widely the concept of regression to the mean can be applied. In financial markets, the best stock pickers find it hard to replicate their success over time. In sports, teams that pay a lot for that breakout star who had a great season are often disappointed. Pretty much anything involving chance or randomness (including height) is going to regress to the mean.

Something else that is not all that important is that Pearson was a genius.

> "Well", [Pearson] said "I do not know how old I was, but I was sitting in a high chair and I was sucking my thumb. Someone told me to stop sucking it and said that unless I did so the thumb would wither away. I put my two thumbs together and looked at them for a long time. 'They look alike to me,' I said to myself. 'I can't see that the thumb I suck is any smaller than the other. I wonder if she could be lying to me'" (Walker, 1958, p. 13).

Even from his high chair, Pearson was skeptical and tested claims with data. Do not believe everything you hear—that is worth remembering.

Figure 4.5: Karl Pearson (1857-1936) generated by ChatGPT.

## References

Walker, H. M. (1958) "The Contributions of Karl Pearson" *Journal of the American Statistical Association*, pp. 11-22; www.jstor.org/stable/2282561.

Figure [illegible] Karl Pearson (1857–1936) [illegible]

## References


[illegible]

# 5. Multiple Regression 3D Visualization

If we have more than one $X$ variable, then we have a multiple regression. Often, we are not really interested in the other included $X$s, but they play a critical role as statistical controls that give us a better understanding of the effect of the $X$ variable of interest on $Y$.

This section shows that multiple regression is an extension of bivariate, $Y$ on $X$, regression. We continue to interpret a multiple regression as a conditional average, but we add the idea of holding constant other included $X$ variables as you interpret a slope coefficient. Understanding that multiple regression controls for other included $X$s is fundamental. Without the complete set of other $X$s added to the regression, the coefficient on an $X$ variable may suffer from *omitted variable bias.*

These ideas are difficult to understand, so we will work first with a simple example and use a 3D scatter plot to develop intuition. Then we will return to our research question and, in a hypothetical example, examine how adding another $X$ variable affects the slope coefficient on education.

## 5.1 The Plane

Fitting a line to an $XY$ scatter plot is relatively easy to understand, but what happens when you add another $X$ variable? Well, you get a 3D (not a 2D) cloud of points and you fit a plane (instead of a line).

In math, the usual notation (which we will adopt) is that the $Z$ axis is the dependent variable and the two axes, $X$ and $Y$, are independent variables. In statistics, however, dependent variables are always $Y$ so we use $X_1$ and $X_2$ as the independent variables.

Suppose we had $Y = 0X_1 + 1X_2$. This equation says that as $X_1$ increases, holding $X_2$ constant at any value, $Y$ stays the same since the slope coefficient on $X_1$ is zero. The slope coefficient on $X_2$ is $+1$ so we know that $Y$ goes up by $+1$ for each unit increase in $X_2$, holding $X_1$ constant.

Plotting this equation produces a plane as shown in Figure 5.1. It is like a flat board tilted up. Notice how the bottom of the plane has the same $Y$ value. This is because as $X_1$ rises (moving from the corner of the three axes towards you), $Y$ stays constant (the slope coefficient on $X_1$ is zero). In fact, this is true for any value of $X_2$ —any vertical slice of the plane for any given value of $X_2$ will yield a horizontal line.

The plane is tilted up because as $X_2$ rises, with $X_1$ held constant, so does $Y$ (the slope coefficient on $X_2$ is 1). The lines in the plane represent vertical slices at given values of $X_1$.

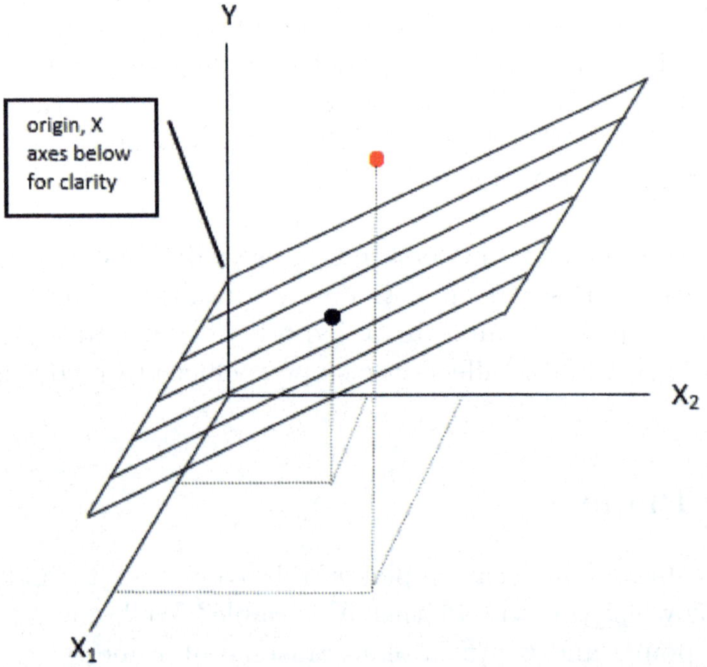


Figure 5.1: The plane given by $Z = 0X + 1Y$ with two example observations.

The two dots in Figure 5.1 represent two individual observations. One dot is on the plane so its residual is zero. The other dot, however, lies above the plane so its residual is positive. Every observation has a residual which is computed by the vertical (on the $Y$ axis) distance from any point to the plane. The plane is fit, like a line is fit, by minimizing the sum of squared residuals.

*STEP* Enter this URL in your favorite browser (or click on the link if you are reading this on a screen) tiny.cc/multreg. If you are logged into Google, you can make a copy of this file to save your work.

Google sheets is a lot like Excel. The data were generated by entering the formula =2*RAND() for both $X_1$ and $X_2$, then copying and pasting as values. $Y$ in cell C2 was generated by the formula =0*A1+1*B1+0.2*(RAND()-0.5). As you will see in a moment, this adds a little random noise to the equation of the plane in Figure 5.1.

We want to regress $Y$ on $X_1$ and $X_2$. This is a multiple regression because we have more than one independent variable. The objective function is the same as before, minimize the sum of squared residuals, but the output will be different. Instead of two, we will get three coefficients, the intercept term and two slopes (one for $X_1$ and another for $X_2$).

Google sheets has a LINEST function, just like Excel. It is actually easier to use because you do not need to select an exactly correct range of cells and then press the *ctrl-shift-enter* keyboard combination. Newer versions of Excel have adopted this feature. It is good to know, however, that some functions are array functions that span across more than one cell.

*STEP* Enter the formula =LINEST(C2:C201,A2:B201,1,1) in cell E2 and press the enter key.

Congratulations! You just ran your first multiple regression. Notice that the coefficients are in the first row and in backwards order, just like in Excel, and the rest of the output is just like Excel's LINEST function.

We want to visualize the data to help us understand what these coefficients mean, but neither Excel nor Google sheets can draw a 3D scatter cloud. We turn to Python, an open-source, high-level, general-purpose programming language.

We can use Python in Google's colab environment to easily visualize the 3D scatter cloud produced by the data we generated.

*STEP* Enter this URL in your favorite browser (or click on the link if you are reading this on a screen) tiny.cc/multregcolab. Be sure to follow the instructions in the colab notebook and spin the charts around. Below, more specific spinning instructions will be provided. For now, just make sure you can spin the plot.

Your spinning should have convinced you that the 200 dots are tightly clustered around a plane. The coefficients from LINEST give us an equation for a fitted plane that minimizes the sum of squared residuals. Notice how close the LINEST coefficients are to the original equation of the plane, $Y = 0X_1 + 1X_2$ (where there is an implicit zero for the intercept). The reason the slope of $X_1$ is not exactly zero and the slope of $X_2$ is not exactly one is because of the noise we added via RAND to help us visualize the 3D scatter cloud.

Next, we turn to interpreting the slope coefficients. We start with the slope of $X_2$ because it is easier to see what is going on. We interpret the slope of $X_2$ as the increase in Predicted $Y$ given an increase in $X_2$, *holding* $X_1$ *constant.* This last part is really important.

Figure 5.2 shows a view of the scatter cloud from the $X_2$ axis side, so we can see the relationship between $X_2$ and $Y$. For any given value of $X_1$ (it could be 1 or 2 or any value of $X_1$), that slice of the cloud will have dots that are roughly on a line with slope one.

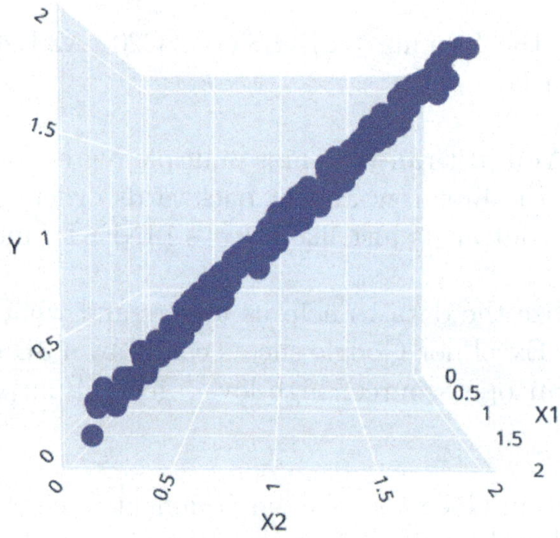


Figure 5.2: A view from the $X_2$ axis.

STEP To replicate Figure 5.2, click the *Home* icon in the chart controls to reset the view to its initial position. Now click the chart and slowly drag right and slightly up until you have the $X_2$ axis at the bottom from 0 on the left to 2 on the right and the $X_1$ axis on the right coming at you from 0 to 2.

When we say "holding X constant," we mean that we are looking at the 3D cloud and fitted plane from the side. If we did not have the noise, the slices of the fitted plane would all be on top of each other and we would see only a line. The fitted plane does not have any noise —remember, it is fitted to minimize the sum of squared residuals and is given by the equation $PredictedY = b_0 + b_1X_1 + b_2X_2$.

Unlike $X_2$, the slope coefficient of $X_1$ is very close to zero. What does that mean? The interpretation is that increasing $X_1$ has no effect on Predicted Y, *holding* $X_2$ *constant.* Again, the language of "holding other included *Xs* constant" is critical.

You are unlikely to have stumbled upon Figure 5.3 as you twirled the chart around. It shows a view of the plane from the $X_1$ axis side.

*STEP* To replicate Figure 5.3, from the Figure 5.2 view, spin the chart horizontally by clicking and dragging slowly to the right.

Figure 5.3 shows why the "holding $X$ constant" language is so important. The fitted plane will be tilted away from you, rising up as it recedes into the background. For any given value of $X_2$ (for example, 0.5, 1, or 1.5), that slice of the fitted plane will be a horizontal line with slope zero.

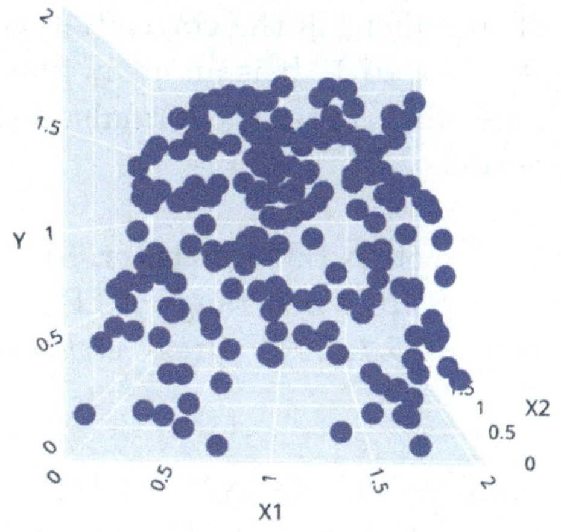


Figure 5.3: A view from the $X_1$ axis.

It is clear that a multiple regression with two $X$ variables produces a fitted plane. What is hard to understand is that when we interpret a slope coefficient in a multiple regression, we are holding the other $X$ constant.

This interpretation and language can be extended when we add even more $X$ variables. We will not be able to visualize the regression as a fitted plane when we have more than two $X$ variables, but we continue to interpret the slope of an individual $X$ variable as its effect on the predicted value of the dependent variable, *holding all other included Xs constant.*

## 5.2 Omitted Variable Bias

We will visualize omitted variable bias by using the same $Y = 0X_1 + 1X_2$ example, with one critical difference —we will construct $X_1$ and $X_2$ so that they are correlated. This is really important.

To do this, we need to have the Monte Carlo simulation Excel add-in installed because it has a user-defined function called BIVARNORMAL. This free add-in is available here: dub.sh/regcourse.

*STEP* Download the *MCSim.xla* file from the link above and use the Add-Ins Manager (File → Options → Add-ins → Go) to install it. Click the *Add-ins* tab and click *MCSim.*

With the add-in available, we can use the BIVARNORMAL array function. It requires five arguments, like this: BIVARNORMAL(avg x1, SD x1, avg x2, SD x2, r). The last argument is the correlation coefficient between the two variables, ranging from -1 to 1. It is an array function, like LINEST, so you must select a cell range first, enter the formula, and press *ctrl-shift-enter* to properly use the function.

*STEP* In a blank Excel workbook, enter the labels $X_1$ and $X_2$ in cells A1 and B1. Next, select the cell range A2:B2 and type the following formula: *=BiVarNormal(5,1,5,1,0.9).* Press *ctrl-shift-enter* to enter this formula.

You should see two numbers in cells A2 and B2. They should both be around 5 give or take 1. But here is the crucial part, they are both highly, positively correlated which means that when one is low so is the other and when one is high so is the other. Critically, $X_1$ and $X_2$ were not connected like this in the first example. This turns out to be really important.

*STEP* Press the F9 key (you might have to use the *fn* key on your keyboard) repeatedly and keep an eye on cells A2 and B2.

Most of the time, when $X_1$ is below 5, $X_2$ is also below 5 and $X_1 > 5$ is paired with $X_2 > 5$. The two numbers are definitely moving together. They are not independent of each other.

*STEP* Select cell range A2:B2 and fill down to row 201. In cell A202, enter the formula *=CORREL(A2:A201,B2:B201)*. Press F9 a few times.

Now you have 200 pairs of correlated numbers and you can see that the correlation coefficient is around 0.9 (as you requested with BIVARNORMAL). It isn't exactly 0.9 because we only have 200 draws. As we increased the number of observations, we would get closer and closer to 0.9.

*STEP* Return to the top of the sheet and enter the label $Y$ in cell C1. In cell C2, enter the formula *=0*A2+B2+0.2*(RAND()-0.5)* and fill it down to cell C201.

The data for the second chart in tiny.cc/multregcolab were generated in exactly this way. Of course, the data were deadened by copying the formulas and pasting them as values, which changed the cells to numbers.

*STEP* Return to colab and scroll down to the bottom chart. Reset to the default view, then apply the two steps (rotate to the $X_2$ view and spin horizontally again so the $X_1$ axis is at the bottom) to get Figure 5.4.

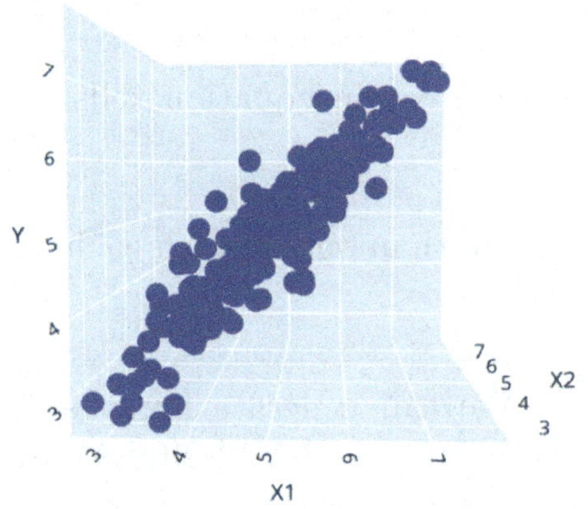


Figure 5.4: A view from the $X_1$ axis when $X_1$ and $X_2$ are highly correlated.

We will compare Figures 5.3 and 5.4, which are quite different, using the same approach: we consider a situation where we ignore $X_2$ in a regression, perhaps because we do not have that variable. This would bring all of the dots forward, pressing them flat at the front of the chart, because there would not be an $X_2$ axis. Removing the $X_2$ axis and bringing all the dots flush at the front produces a 2D scatter plot in $X_1, Y$ space.

The key idea is that there is no relationship between $X_1$ and $Y$ in Figure 5.3 because the dots in that 2D scatter would be a formless blob. They would spread all over the 2D plot and the resulting fitted line of $Y$ on $X_1$ would have a slope of zero. This is a good result because the slope coefficient on $X_1$ actually is zero.

But now look at Figure 5.4 and perform the same thought experiment. Suppose we ignore $X_2$ and regress $Y$ on only $X_1$, what happens? The dots all come forward, but there is a clear association between $Y$ and $X_1$. Fitting a line to the 2D scatter plot produced by ignoring $X_2$ in Figure 5.4 will give us a positive slope coefficient on $X_1$. This is omitted variable bias. We will think that $X_1$ affects $Y$, when it does not.

The cause of the difference in Figures 5.3 and 5.4 is the correlation between $X_1$ and $X_2$. When the correlation is zero, we do not have any omitted variable bias. When it is not zero, we do.

The next chapter will make this point again, but it will focus on the regression coefficients. Remember Figures 5.3 and 5.4 as you run regressions.

## 5.3 Takeaways

The slope of a bivariate regression equation tells us how a change in $X_1$ affects *Predicted Y*.

Multiple regression has more than one $X$, with each one having its own slope coefficient.

> *EXCEL TIP* We can generate hypothetical data in Excel for a multiple regression by using the BIVARNORMAL array function. The MCSim add-in must be installed and it requires five arguments:
>
> =BIVARNORMAL(avg x1, SD x1, avg x2, SD x2, r)

We often include other $Xs$ in the regression, not because we are interested in their effect on $Y$, but because they are controls. They enable us to interpret the coefficient on the $X$ variable of interest conditional on the other included $Xs$ remaining constant.

If we omit an $X$ variable, it might impact the coefficient of the included $X$ variable in the regression. In terms of the 3D visualization, omitting an X variable removes its axis and you are left with a conventional 2D scatter instead of a 3D cloud of points.

If the two Xs are independent (unconnected and uncorrelated), then the included X remains essentially unaffected. In Figure 5.3, there is no relationship between $X_1$ and $Y$ even if we suppress the $X_2$ axis.

However, if we are in the usual situation where the Xs are intertwined, then we are in trouble if we omit one of them. In Figure 5.4, the 2D scatter plot is clearly depicting a positive relationship between $X_1$ and $Y$ even though the true coefficient on $X_1$ in the multiple regression is zero. This is omitted variable bias.

# 6. Omitted Variable Bias

The previous section used 2D and 3D visualizations to show that omitted variable bias, OVB, occurs when an omitted $X$ is correlated with an included $X$. This section continues focusing on omitted variable bias, but does so by comparing the slope coefficients of regressions. This will show exactly what we mean by omitted variable bias.

After describing and demonstrating the omitted variable bias rule, we will construct a hypothetical data set calibrated to the regression of INCWAGE on EducYears. Again, you will see concrete numbers reflecting omitted variable bias.

Finally, we end this section with a series of intuitive examples of omitted variable bias. They will help us understand how observational studies are fundamentally different from, and inferior to, controlled experiments.

## 6.1 The Omitted Variable Bias Rule

Consider these three regressions with data created from the equation $Y = 0X_1 + 1X_2 + 0.2 * (RAND() - 0.5)$:

1. Y on $X_1$ and $X_2$ (the correct specification)
   $PredictedY = b_0 + b_1X_1 + b_2X_2$

2. Y on $X_1$ (the misspecified model with an omitted variable; $g_1$ is biased)
   $PredictedY = g_0 + g_1X_1$

3. $X_2$ on $X_1$ (an auxiliary regression to help explain omitted variable bias):
   $PredictedX_2 = d_0 + d_1X_1$

We will demonstrate, but not derive, the *Omitted Variable Bias Rule*:

$$g_1 = b_1 + d_1b_2$$

This equation says that the slope coefficient, $g_1$, on the misspecified regression (with the omitted variable $X_2$) is biased by an amount equal to the slope coefficient, $d_1$ of the auxiliary regression times the slope coefficient, $b_2$ of the omitted variable in the correct model. We will show this to be true with data from the previous section.

We have data from two different situations with which we can test the three regressions to help us understand the meaning of omitted variable bias. Both datasets were built with same equation, $Y = 0X_1 + 1X_2 + 0.2 * (RAND() - 0.5)$, but one has the two *Xs* unrelated to each other and the other used BIVARNORMAL with $r = 0.9$.

We begin with a multiple regression with the *correlated Xs.* If you do not have these data, return to the previous section and generate 200 observations with $Y, X_1, X_2$ variables by following the instructions for using BIVARNORMAL.

*STEP* Run a regression with the correct *specification* (the functional form and included *Xs* in a regression), which is Y on $X_1$ and $X_2$. Press F9 a few times.

It is clear that the intercept is around zero, with the slope of $X_1$ also around zero, but the slope of $X_2$ is around one. These coefficients are not exactly 0, 0, and 1 because of the random noise we added. Notice also that the SSR is quite low compared to the RSS, so the $R^2$ is really high. Finally, the RMSE is quite low. These are all indicators of excellent results for our regression.

But what if we did not know that $Y$ depended on $X_1$ and $X_2$? What if we thought $X_1$ was the only variable that mattered, therefore, we regressed $Y$ on $X_1$ alone?

*STEP* Regress $Y$ on $X_1$ alone. You have omitted $X_2$.

The problem is immediately clear. You are staring directly at omitted variable bias. The slope coefficient is wrong. This all started with $Y = 0X_1 + 1X_2$ so $X_1$ has no effect on $Y$, yet the slope coefficient in your regression of $Y$ on $X_1$ (alone) is nowhere near zero. We say that the slope coefficient is *biased* because it is not bouncing around its true value, zero. The cause of this deviation from the true value lies in the omission of $X_2$ so it is called omitted variable bias.

Notice also that there are no red flags in any of the typical regression diagnostics. The $R^2$ is pretty good and the RMSE is reasonable. All looks well. There is no test and no indicator of omitted variable bias. It can only be discovered by critical thinking about the research question and the data generation process.

No matter how many times you press F9, the slope coefficient remains very far from zero. The bivariate regression makes it seem like $X_1$ affects $Y$, but we know this is not true.

We can run another regression to help us understand the source of the bias. This regression will use the omitted $X_2$ variable as an artificial dependent variable. This regression will enable us to measure the connection between $X_1$ and $X_2$. We will see that this connection is absolutely critical in generating omitted variable bias.

*STEP* Use LINEST to regress $X_2$ on $X_1$. The formula for LINEST is: *=LINEST(B2:B201,A2:A201,1,1)*. Press F9 a few times and keep your eye on the slope coefficient.

The slope coefficient on this auxiliary regression, $d_1$, conveys the connection between $X_1$ and $X_2$. We avoid omitted variable bias if either $d_1$ or $b_2$ are zero. We know $b_2$ is one and we now have evidence that $d_1$ is not zero, therefore, we have omitted variable bias.

*STEP* In a blank cell compute $g_1$ (the slope coefficient in $Y$ on $X_1$) as $b_1$ (the slope on $X_1$ in the first, correct multiple regression) plus $d_1$ (the slope from the auxiliary regression) times $b_2$ (the slope on $X_2$ in the first, correct multiple regression).

Wow! This cell that you just created is exactly equal to the biased slope in the regression of Y on $X_1$. In fact, the amount of the bias in $g_1$ is $d_1 b_2$.

Figure 6.1 shows how the Omitted Variable Bias Rule combines the correctly specified and auxiliary regressions to exactly replicate the slope of the misspecified regression.

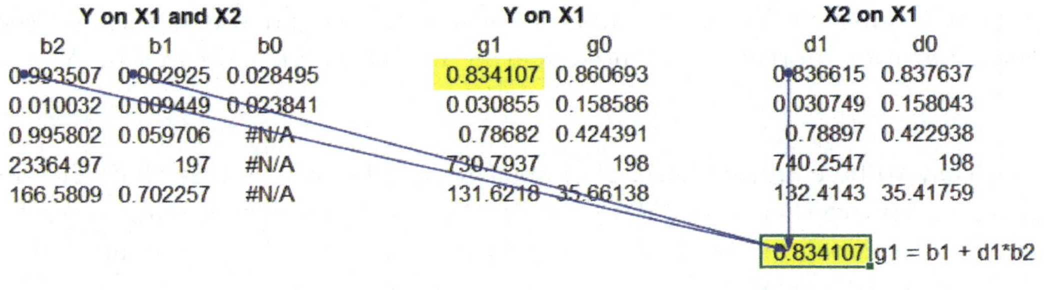


Figure 6.1: Applying the Omitted Variable Bias Rule

Your numbers will be a little different because you have live data that is slightly changing every time your press F9, but Figure 6.1 should help you compute the key result that the two yellow-highlighted cells are exactly equal. The arrows were produced by Excel's *Trace Precedents* (in the *Formula* tab in the *Ribbon*).

Figure 6.2 helps show that the reason why the slope in $Y$ on $X_1$ ($g_1$) alone is biased is because there is a connection between $X_1$ and $X_2$ (captured by $d_1$) and $X_2$ affects $Y$ (captured by $b_2$). When you omit $X_2$, it is still there, lurking underneath, as captured by the hatched pattern over $X_2$ in Figure 6.2.

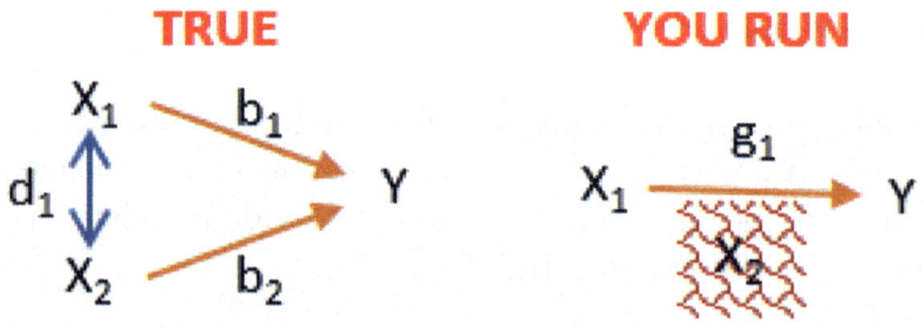


Figure 6.2: Depicting the Omitted Variable Bias Rule

As $X_1$ increases in the regression of $Y$ on $X_1$, it is true that Predicted $Y$ increases by $g_1$, but $X_1$ is not generating the increase by itself. Because $X_1$ and $X_2$ are connected, increasing $X_1$ is also increasing $X_2$ and it is $X_2$ that is increasing $Y$. $X_1$ is "getting the credit" in the misspecified, bivariate regression for the effect that $X_2$ has on Y.

We saw this effect this with our 3D plot visualization in the previous section. If you regress $Y$ on $X_1$ and $X_2$, the slope on $X_1$ is correctly zero because you get a slice from the plane for each value of $X_2$. But if you omit $X_2$, and regress $Y$ only on $X_1$, the $X_2$ axis would disappear and you would just see that southwest-northeast oriented scatter in 2D where increases in $X - 1$ increase $Y$.

It is easy to be confused here. Perhaps this will help: it is true that $X_1$ and $Y$ move up together in our data set ($g_1$), but not because $X_1$ causes $Y$. $X_1$ and $Y$ are associated because higher $X_1$ leads to higher $X_2$ ($d_1$) and higher $X_2$ leads to higher $Y$ ($b_2$).

If we want to know the isolated, own effect of $X_1$, we need to hold $X_2$ constant. We can do that if we have the correctly specified, full model.

There are two ways to not get OVB. The case where $b_2 = 0$ is trivial. It says that we will not get omitted variable bias when we ignore $X_2$ if $X_2$ does not matter. The other case occurs when $X_1$ and $X_2$ are independent. That was the situation in our first data set in the previous section.

*STEP* Return to the Google sheet RAND in tiny.cc/multreg and run three regressions to demonstrate that the OVB Rule works. Explain why the bias is so low.

## 6.2 Hypothetical Drive Variable

We practice with the OVB Rule by returning to our research question, what is the rate of return to education? We use a pretend version of the 2024 CPS data set.

We add a new variable, not in the CPS and completely made up, so everything below is *fake data* designed to explain omitted variable bias.

*STEP* Open the data set in Google sheets by going to tiny.cc/IncEducDrive.

There are 50,124 observations with INCWAGE and EducYears defined as before. The new, hypothetical variable is *Drive.* It is a combination of innate intelligence, motivation, charisma, reliability, grit, and social skills. It affects income, just like education, more of it produces higher income.

We cannot ask people how much drive they have because they would not know. Everyone has some drive, but some people have more than others. Unfortunately, it is not directly measurable. People with more of this magical drive stuff earn more income because their higher drive makes them more productive workers.

*STEP* Make a histogram of Drive. In Google sheets, *Chart* is listed under *Insert.* The default is a column chart, but Google sheets has a histogram option in the *Chart Editor* dialog box on the right.

Your histogram shows that Drive ranges from 80 to 120 and is roughly normally distributed, with min 83, max 116, and the average is 99.

The higher Drive, the higher INCWAGE, but we are not interested in this. We want to know how much more education increases income. We have an answer from previous work, \$11,073/year. That was the slope coefficient on a regression of INCWAGE on EducYears. Is this still true in this data set?

*STEP* Use LINEST to regress INCWAGE on EducYears.

Yes, the slope coefficient on this regression is roughly \$11,073, just like in the real CPS data set. This is $g_1$, the slope coefficient in the misspecified model.

The slope coefficient on EducYears is biased because Drive was not included in the regression, Drive affects income, and Drive is connected to EducYears.

*STEP* Compute the correlation coefficient between Drive and EducYears. What does it tell you?

You found that r = 0.92, that is a high positive association. When EducYears is high, Drive is also likely to be high and vice versa.

The correlation coefficient conveys information about the relationship between Drive and EducYears, but to apply the OVB Rule, we need to run the auxiliary regression.

*STEP* Make a Scatter chart of Drive as a function of EducYears. Fit a Trendline and display the equation of the fitted line along with the $R^2$. Run LINEST to confirm the Trendline results.

You found that the slope coefficient on EducYears in this auxiliary regression is 2. This confirms what $r$ told us, these two variables are connected. We now have $d_1$ in the OVB Rule framework. We still need $b_1$ and $b_2$ to use the OVB Rule.

*STEP* Use LINEST to regress INCWAGE on EducYears and Drive. This is, hypothetically speaking of course, the correctly specified model.

You found that $b_1$ is 8075 and $b_2$ is 1501. Did the OVB Rule work? Yes, we simply need to plug in the values to see that it did:

$$g_1 = b_1 + d_1 b_2$$
$$11073 = 8075 + 2 \times 1501$$

The small difference is due to rounding error. Use the values in the LINEST cells if you want to show that the OVB Rule exactly works.

This work shows that our original estimate of the value of one more year of schooling, \$11,000 was off by a lot. One more year of education gets you around \$8,000, *holding Drive constant.*

But two things. First, we made up the Drive variable. We do not actually know the true answer to the amount of omitted variable bias caused by the omission of a measure of "drive." And second, even we did have that variable, notice that the $R^2$ has improved quite a bit, but we are still far from the correct model. In other words, there are many other missing variables so we still have omitted variable bias in EducYears.

It is true, however, that the regression analysis with Drive is better than it is when we ignore this effect. We cannot get a perfect answer, but there are better and worse answers.

## 6.3 Other Examples of OVB

Suppose you want to know how much one more hour studying each week affects grades. Suppose you (somehow) have data from many students on hours per week studying and GPA (grade point average on a scale where an A is a 4.0). You regress GPA on hours and get a slope coefficient of say 0.05. This says that studying one more hour per week gives an increase of 0.05 Predicted GPA.

Is this a good answer to the question? No, absolutely not. The problem is that we are ignoring many other factors correlated with hours studying that also affect grades. What if smarter students study more? Then we will be tricked into thinking that studying is more valuable than it really is. When a variable's effect is incorrectly estimated because other effects are not accounted for, we say there is *omitted variable bias*, more colloquially known as *confounding.*

If we somehow measure student ability (maybe with standardized test scores or high school grades) and add it to the regression, then when we interpret the coefficient on hours we can say, "holding student quality constant." This is better analysis, but still not perfect.

Suppose you want to know how much more exercise affects health. Suppose you (somehow) have data on individuals that includes how many hours each person exercises per week and how healthy they are measured on a scale from 0 to 10. You regress health on exercise and the slope coefficient is 1.0. One more hour pew week of exercise increases health status by +1.

Is this a good answer to the question? No, absolutely not. The regression fails to hold anything constant. Surely there are more factors affecting health and they are likely to be correlated with exercise. For example, people who exercise more are likely to have better diets. Leaving out the omitted variables from the regression confounds the slope coefficient on exercise. It is a combination of the effect of exercise on health *and* the effect of other variables on health that are connected to health.

You know this one, but just to be sure: suppose you want to know how education affects income. You get CPS data and regress INCWAGE on EducYears. The slope coefficient on EducYears is roughly $11,000 per year. One more year of education produces, on average, $11,000 per year more income so going to college is worth about $44,000 per year more than a high school degree.

Is this a good answer to the question? No, absolutely not. The slope coefficient suffers from omitted variable bias. It is getting too much credit. For example, it is likely that intelligence, ambition, experience—believe it or not, even height —and many other factors affect earnings *and* these factors are correlated with educational attainment. By ignoring these factors, the EducYears variable sops up those effects and it appears to be more important than it really is.

Another way to understand the problem of omitted variable bias is to consider this scenario: what if we plucked a high school grad out of the data set and magically gave them a college degree? Would they make $44,000 per year more? No, because there are many other factors that we did not change. If we held constant all of the other variables that affect income, then the coefficient on education would tell us how much more income more education would give.

## 6.4 Another Strategy to Avoid OVB

Perhaps you have heard that randomized experiments are the gold standard in research. They take a different approach to the problem of omitted variable bias. Instead of statistically controlling for other variables, they employ *random assignment* to prevent OVB.

Medical testing uses this randomized controlled trial (RCT) strategy. In essence, you split participants into experimental and control groups. The former gets the new drug and the latter get a placebo. You see if there is a difference in health between the two groups. There are many complications (both groups have to adhere to the protocol), but this works by randomly assigning the other variables that matter.

For the effect of studying, the experiment would be something like recruit students, randomly assign them a number of hours to study, then compare their grades. You might be smart and be told you can only study 2 hours in a week. You do not get to choose how much to study. This breaks the connection between other factors and hours studying.

For the effect of exercise, recruit people and randomly assign them to exercise a certain number of hours each week then see how healthy they are. The random assignment cancels out "the healthier people exercise more" bias. After all, healthy people are assigned to high, medium, and low amounts of exercise. They do not get to choose.

Neither of these two experiments would pass IRB (Internal Review Board) review. And you are certainly not going to randomly assign 18 year-olds to go to college or work, then wait decades to see how they do in the labor market.

We are often faced with data from *observational studies*, where people decide what to do. We can ask questions or observe them, but we cannot randomly assign them to block the effects of omitted variable bias. Our only hope is to include as many other factors as we can identify in our regression.

Although outside our scope, there have been advances in random assignment such as A/B testing and natural experiments. This speaks to the powerful advantages that randomized methods have over observational studies.

## 6.5 Takeaways

You have heard that "correlation is not causation." That is another way of understanding the danger of omitted variable bias. Simply examining the relationship between two variables and drawing conclusions about the effect of one variable on another is almost always a mistake.

Adding $X$ variables to a regression changes the coefficients of previously included *Xs*. This means we have to decide which specification (the functional form and included explanatory variables) is the correct one.

When we do not have the model perfectly specified, we worry about omitted variable bias. This means that we are misled and misunderstand the effect of an explanatory on a dependent variable.

With $b_2 \neq 0$, the usual case, $d_1$ is critical. The only way to get $d_1 = 0$, which ensures that $g_1 = b_1$ and there is no omitted variable bias, is for $X_1$ and $X_2$ to be independent. If the two Xs are connected, we have $g_1$ suffering from OVB.

There is no test for omitted variable bias.

The OVB Rule, $g_1 = b_1 + d_1 b_2$, is impractical. In the real world, we never know $d_1$ or $b_2$. The OVB Rule is just a way to help us understand how omitted variable bias affects a slope coefficient.

A randomized controlled experiment protects against omitted variable bias. The randomization ensures that no omitted *Xs* are varying systematically with the variable we are interested in.

Unlike observational studies where it is almost guaranteed that omitted *Xs* are connected with included *Xs*, the random assignment of a controlled experiment produces unconnected omitted Xs.

In our Y, $X_1, X_2$ model, a controlled experiment means $d_1$ (in the auxiliary regression) is probably close to zero. We will see exactly how random assignment fixes things in a future model.

Although we never have the perfect model and we always have omitted variable bias in an observational study, do not make perfect the enemy of good. We can do better analysis if we have some controls. This may be the most important takeaway of all.

# 7. Functional Form

The goal is to understand that even though it is called linear regression, it can describe non-linear relationships in the data. In other words, a linear regression can produce curved fits.

The key is the fact that linear regression means "linear in the parameters," not in the variables. This means that regression is incredibly flexible—certainly much more flexible than most people think.

We illustrate this claim with two examples, one with hypothetical data with a non-linear X and the other with our CPS data with a natural log transformation of income.

## 7.1 Non-linear X

Galileo famously showed that the relationship between time and the distance an object falls is non-linear. We will generate artificial data based on Galileo's work.

Our $X$ variable is the *time*, $t$, in seconds at which we measure the distance an object has fallen from the roof of a hypothetical tall building.

*STEP* In cell A1, enter the label *time* and in the ten cells below enter the series 1 to 10 by 1.

The dependent, or $Y$ variable, is *distance*, $d$, in feet, given by the equation $d = 16t^2 + \epsilon$, where epsilon is a random measurement error. We will use Excel's RAND() function for epsilon.

*STEP* In cell B1, enter the label *distance* and in cell B2, enter the formula *= 16*A2^2 + RAND()*. Fill it down.

We want a single sample that does not keep bouncing around as we work on the sheet, so we need to deaden the cells.

*STEP* Select cells B2:B11, copy, and then paste values (click the down arrow in the *Paste* button and select *Paste Values*) to replace the formulas with numbers.

If we draw a scatterplot, we can easily see that distance and time have a non-linear relationship. The object is accelerating as it falls (the derivative with respect to t is 32 feet per second squared) so the longer you give it to fall, the distance increases at an increasing rate (until you reach terminal velocity, of course).

*STEP* Create a chart of *distance* as a function of *time*, then add a trendline and display the equation on the chart, along with $R^2$.

It is pretty obvious that the regression does not describe the data well. Notice that the high $R^2$ is irrelevant. This regression is no good because a straight line is a poor description of the curved relationship between distance and time.

Can linear regression handle this situation? Yes, it can. Even though it is called linear regression, it can produce curved fits. This language is confusing—how can something linear produce a curve?

We begin with a new $X$ variable, *time squared.*

*STEP* Insert a column between time and distance and label it *time2.* Select the 2 in the formula bar and make it a superscript so the cell displays "time$^2$". In cell B2, enter the formula *=A2^2* and fill it down.

We can now regress *distance* on *time squared.*

*STEP* Create a chart of *distance* on *time squared* and add a trendline, displaying the equation and $R^2$.

The result is spectacular. Regression is now working really well. Note that, although the $R^2$ shows as 1, it is not exactly 1 (a perfect fit). This is because of the noise we added with RAND.

*STEP* Run LINEST on *distance* and *time squared* to see that $R^2$ is extremely close to, but not exactly 1.

We return to the confusing issue of the phrase *linear regression.* Is this a non-linear regression? No. We still have an equation in the form of a line, $y = mx + b$. Even though we **transformed** the $X$ variable (by squaring time), the functional form is still said to be "linear in the parameters."

We can do all kinds of transformations to Y and X, but as long as we maintain the $y = mx + b$ form, it is a linear regression.

There is such a thing as non-linear regression, but it involves non-linearity in the coefficients. For example, $y = b_0 + \frac{1}{b_1}x$ would be a non-linear regression. This is beyond the scope of this book.

The advantage of transforming an $X$ or $Y$ variable is that we can use linear regression to produce all kinds of curved fits. The next example shows this for $Y$.

## 7.2 The Earnings Function Theory

As we saw in previous work, fitting a line to a scatterplot of INCWAGE and EducYears gives poor results. The CMF is clearly curved so regressing income on years of schooling is similar to regressing distance on time. We need to capture the non-linearity in the relationship.

Why is this happening—why is average income as a function of education rising at an increasing rate? Because education is an investment and we have to present value future income. This produces higher proportional returns so one more year of education increases income by a *percentage*, not by a dollar amount.

Can we fit a curve with linear regression? Yes, by applying the appropriate transformation to INCWAGE—we will take the natural log of INCWAGE and regress it on education. This is called a *semi-log functional form*: $\ln Y = b_0 + b_1 X$. It is well-named since only the left half of the function is transformed.

We begin with a theoretical model of education that is known as *human capital theory*. It treats humans as machines who invest in themselves to increase their productivity and, thus, income. Crucially, income is a flow over time. After investing, you get higher income for many years in the future.

*STEP* In cell A1, enter the label *Year* and then create a series from 0 to 39 in cells A2:A41.

This is a person's 40-year working lifespan in our toy model. Year 0 represents the present, current year.

There are two jobs available, Job 1 requires no skills or knowledge so anyone can do it and pays $30,000 per year. Job 2 requires one year of training or schooling and, currently pays $32,000 per year. The two jobs are identical in every other respect.

Training for Job 2 has no out-of-pocket cost, but it does have an opportunity cost of foregone income. You cannot work while being trained so you are giving away $30,000 of income when you decide on Job 2.

*STEP* Create *Job 1* and *Job 2* labels in cells B1 and C1, respectively. Enter $30000 in cell B2 and fill it down. In cell C2, enter *$0* and in cell C3 enter *$32000.* In cell C4 enter the formula = *C3* and fill it down.

Which job would you take? You might notice that the total dollars of Job 2 are higher so Job 2 is the better choice? No, this is incorrect.

Because the dollars come at different points in time, they are not in the same units and cannot be summed. Excel, of course, is happy to add the numbers in cells B2 to B41, but this is an illegal sum because the numbers have different units. It would be like adding 3 cars and 4 pencils to get 7 carpencils.

We have to *present value* the dollars in the future years so they all have the same time units. The present value is the current value of a future amount given a specified rate of return. The present value of $1 $n$ time periods ($t$) in the future is

$$PresentValue = \frac{1}{(1+r)^t}$$

The *discount rate*, $r$, is the rate of return for money invested today. Let's suppose it is 10%.

*STEP* In cell D1, enter *10%* and the label *discount rate* in cell E1. Name cell D1, *r_*. In cells F1 and G1, enter the labels *PV Job 1* and *PV Job 2*, respectively. In cell F2, enter the formula = *B2/(1+r_)^A2* and fill it down. In cell G2, enter the formula = *C2/(1+r_)^A2* and fill it down. Format the cells as dollars with no cents for legibility.

Notice how the dollars farther in future get smaller and smaller. This is because it takes fewer dollars invested today to grow into the same target dollar amount farther in the future.

Most importantly, the dollars in columns F and G share the same time unit—the present or year 0. By discounting the dollars in the future so they share the same unit of time, we can add them correctly.

*STEP* In cell F42, enter the formula *=SUM(F2:F41)* and in cell G42 enter the formula *=SUM(G2:G41)*. Select cells F41 and G41 and format them with a bottom double border (in the *Font* group of the *Home* tab). This is an accounting convention to show the cells are being totaled.

Our present value work shows that Job 2 pays more, but not enough to make it worth the initial investment of foregoing $30,000 in the first year. Since the present value of Job 1 is higher than Job 2, the higher pay of Job 2 is insufficient to attract enough workers.

*STEP* In cell H42 enter = *G42–F42*.

Because H42 is negative, we know Job 1 is the better choice. If H42 was positive, Job 2 would be chosen. We can use the value of cell H42 as an easy way to decide which job to take.

If no one is attracted to Job 2, the pay differential between the two jobs must rise. What if employers of Job 2 offered $35000—would that be enough to entice people to acquire the training needed for Job 2?

*STEP* Enter 35000 in cell C3.

Cell H42 is now positive so Job 2 is the better choice and workers start lining up for Job 2. But this salary gap is too big. Where is the equilibrium differential? We need to find the gap that makes workers indifferent between the two jobs.

*STEP* Click the *Data* tab and click *Solver*. Cell H42 is our objective and we want the value to equal zero by choosing cell C3. Click Solve.

With a discount rate of 10% and Job 1 paying $30,000 per year, Job 2's equilibrium salary is $33,075. This makes the present values of the two jobs equal. This means that the $3,075 differential is exactly the amount needed to make a new person deciding which way to go indifferent between the two jobs. We need to record this result.

STEP In cell I1, enter the label *Years of training* and in cell J1 enter the label *income*. In cell I2 enter a 0 and in cell J2 enter $30000. In cell I3 enter a 1 and in cell J3 copy and paste the value from cell C3.

There is another way to decide whether or not to make the investment in the training needed for Job 2. We can compute the *internal rate of return*, IRR, of the investment and compare it to the discount rate. If the IRR is greater than discount rate, then we make the investment.

STEP In cell Q1, enter the label *net income* and in cell Q2 enter the formula *=C2–B2*. Fill it down.

This shows the investment project. The up front cost of $30,000 (a negative number) is followed by a stream of payments (positive numbers). The IRR is the interest rate that makes the net present value of all the flows equal to zero.

STEP In cell R1, enter the formula *=IRR(Q2:Q41)* and hit enter.

Excel reports that this project has a rate of return of 10% which equals the discount rate. Thus, just like the present value method, the IRR method is saying we are indifferent between the two jobs.

What if Job 2 required two years of training? What would be the income required to make someone be indifferent between the two jobs?

STEP Change cell C3 to 0 and cell C4 to 33075. Is this salary enough to attract workers to a Job 2 that requires two years of foregone income?

No way. Cells H42 (the difference of the sum of present valued dollars) and R1 (the internal rate of return) both agree that giving up two years is going to require employers of Job 2 to pay more than that. How much more?

STEP Run Solver again, setting the changing cell to C4.

The equilibrium income for Job 2 with two years of training is $36,473.

STEP Add this result to the table.

We can immediately see that income is rising faster than by a constant dollar amount. For one year of training, the income bonus was $3,075, but for two years of training, it has increased by more than that.

**STEP** Find the equilibrium income levels for Job 2 with 3 and 4 years of training.

How is income rising as a function of years of training? Is it by a constant dollar amount?

**STEP** In cell K1, enter the label *Dincome* and then select the letter *D* and make it Symbol font so it displays as the Greek letter delta, Δ. In cell K3, enter the formula *=J3–J2* and fill it down.

This is definitive proof that income is not changing at a constant dollar amount as training rises. Income is rising at an increasing rate as the years of training increase.

**STEP** In cell L1, enter the label *%Dincome* (change the *D* to Δ) and in cell L3, enter the formula *=(J3–J2)/J2* and fill it down.

Wow! This shows us that income is following an exponential path. Equilibrium income differentials for jobs requiring more training are constant *multiples* of each other.

We can linearize this relationship by taking the natural log of income.

**STEP** In cell M1 enter the label *ln income* and in cell M2 enter the formula *=LN(J2)*. Fill it down. Create a chart of ln income as a function of years of training and add a trendline, displaying the equation and $R^2$.

The points are almost exactly on a line and the slope coefficient is extremely close to the rate of return on training. This regression, the natural log of income (or any measure of earnings) as a function of education (or training), is called an *earnings function.*

It is a semi-log functional form because we take the ln of $Y$, but leave the $X$s in their own units. There is a double-log functional form where we take the logs of both $Y$ and $X$ variables.

The slope coefficient is slightly off for two reasons. First, the coefficient itself is an approximation to the percentage change. To get the percentage change from a semi-log regression exactly right, we have to compute $e^{slope} - 1$. But even this will not give us exactly 10% because our example is based on a finite number of time periods. The more years we add, the closer we get to 10%.

The fundamental lesson is that the natural log transformation linearizes the curved, exponential relationship between income and education. The source of the curvature is found in the present valuing of the benefits of education coming in the future. To make the investment in education worthwhile, pay must rise exponentially as education rises.

## 7.3 The Earnings Function in Practice

How can we apply this lesson to our real-world CPS data set? That is easy—we need to transform INCWAGE by taking the natural log before we regress it on education.

*STEP* Open your CPS data set and insert a column between INCWAGE and EducYears. Label it *ln INCWAGE* and enter a formula in the cell below that uses Excel's LN function to take the natural log of INCWAGE. Fill it down.

Now for the real test. IF we regress ln INCWAGE on EducYears, what effect will this have?

*STEP* Copy the existing scatter chart and paste it on the sheet, then edit the SERIES formula, replacing INCWAGE with ln INCWAGE. Do the same with LINEST. Copy the LINEST formula, select a 5 x 2 range of cells, then edit the formula to regress ln INCWAGE on EducYears.

We cannot directly compare $R^2$ (or RMSE or SSR) values when you transform the $Y$ variable. To see that the semi-log regression is doing much better, we can examine the CMF.

*STEP* Copy the PivotTable and paste it, then replace INCWAGE with ln INCWAGE. Copy the new PivotTable and paste it, then add a column for Predicted ln INCWAGE and use the LINEST coefficients to predict ln INCWAGE given years of schooling. Finally, copy and paste the chart and edit the SERIES formula to show the CMF and regression line with the ln INCWAGE data.

Figure 7.1 compares the INCWAGE and ln INCWAGE regressions. The results are eye-opening. The CMF for the ln INCWAGE version (bottom panel) is much more linear and the regression is giving a much better fit.

Unlike the regression with INCWAGE, where the slope coefficient was easy to interpret as the increase in dollars of income for each year of education, we have some work to do to interpret the slope coefficient with ln INCWAGE. The slope coefficient in an earnings function tells us the average percentage return to one more year of education. It is often used by itself as an approximation, so our regression says the return to education is about 11.7%. The exact answer, however, requires that we compute the antilog.

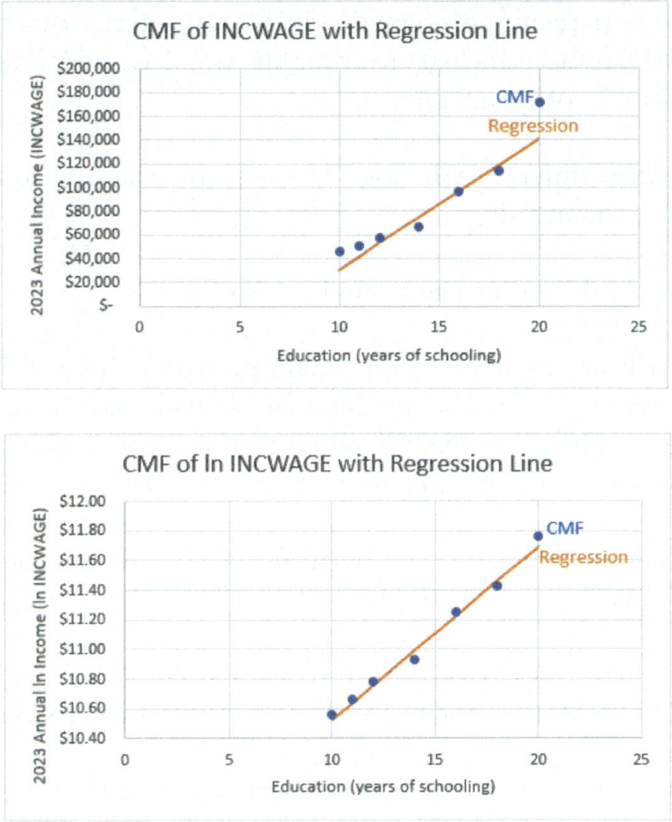


Figure 7.1: The Effect of ln INCWAGE.

STEP In a cell below LINEST in the data sheet, enter the formula *=EXP(cell with slope)−1.* Thus, our regression tells us that income increases by 12.4%, on average, for each year of additional education. An explanation of the algebra behind this is in the appendix.

Why is the coefficient of an X variable with ln Y a percentage change? Because the change in ln Y is approximately equal to the percentage change in Y. A simple numerical example will convince you.

STEP Insert a blank sheet and enter 200 in cell A1 and 201 in cell A2. In cell B2, compute the percentage change from 200 to 201 by entering the formula *=(A2−A1)/A1.* In cell C1 enter *=LN(A1)* and in cell C2 enter *=LN(A2).* In cell D2 enter *= C2−C1.* Compare cells B2 and D2.

The difference in the natural logs (in D2) is very close to the percentage change in cell B2. This is why a slope coefficient on X with ln Y is interpreted as a percentage change. It is a property of the natural log function that differences in logs are roughly equal to percentage differences in the original units of a variable.

We can make the difference in logs give us the exactly equal percentage change by using the anti-log.

STEP In cell E2 enter the formula *=EXP(D2)−1.*

Cells E2 and B2 are exactly equal, 0.005 or 0.5%. The difference in logs is approximately equal to the percentage change and the anti-log of the difference in logs minus 1 is exactly equal to the percentage change. Be sure to read the appendix if you want to see the math behind this.

There is another benefit from the natural log transformation as applied to regressions of earnings on education.

STEP Add the SD of INCWAGE to the INCWAGE PivotTable and add the SD on ln INCWAGE to the ln INCWAGE PivotTable. Compare the SDs.

The key point is not that the natural log has reduced the SDs, but that the SDs are much more uniform. Thus, the natural log transformation helps reduce the *heteroskedasticity*, which means different spread, inherent in earnings distributions. We will see in future work that the presence of heteroskedasticity causes trouble for our estimate of variability in the regression coefficients.

## 7.4 Takeaways

Linear regression means linear in the parameters. We can make all kinds of transformations of the $y$ and $x$ variables, but as long as we maintain the $y = mx + b$ structure, we are doing linear regression.

These transformations allow us to fit all kinds of curves and non-linear shapes while still doing linear regression.

Perhaps the most important transformation is applying the natural log to a measure of earnings and regressing it on education (and other $Xs$). This semi-log functional form is called an *earnings function* and is the most often performed regression in the history of economics.

Excel, and everyone else, abbreviates the natural log as *ln* (instead of the seemingly more obvious *nl*) because it is *logarithmus naturalis* in Latin. This is good for Jeopardy! You will have to search for yourself as to why it is a "natural" log.

In earlier work, we found that one more year of education was associated with about \$11,000 of additional income each year. With the natural log transformation, we find that an additional year of schooling increases annual income by about 13%, on average. Both of those numbers are quite high, however, we must remember that we have not controlled for any other factors. Education is getting too much credit because there are many other factors that affect income that are correlated with education.

The language used for the coefficient on education in an earnings function is pretty loose. To compute a true rate of return, in the financial sense of a discount rate that sets the net present value to zero, we would have to know the out-of-pocket costs of education (such as tuition and books). We obviously do not have that information from the CPS, but we often interpret the coefficient on education in a semi-log earnings function as a rate of return to education anyways.

The primary reason why we take the natural log of earnings was discovered by Jacob Mincer in the 1950s. Mincer (1958) showed that pay differentials would be multiplicative instead of additive for jobs with different amounts of training or schooling.

A secondary reason for taking the natural log of earnings is that it makes the spread in earnings and the resulting residuals more uniform. In other words, it helps reduce the heteroskedasticity in the data. This is beneficial, although we cannot explain why until we deal with the fact that we are working with a random sample from a population.

Figure 7.2: Jacob Mincer (1922-2006) generated by ChatGPT.

The earnings function is one of the best examples of how theory is used to determine the functional form, or specification, of the regression equation. It has deep theoretical foundations, going all the way back to Adam Smith.

> The whole of the advantages and disadvantages of the different employments of labour and stock must, in the same neighbourhood, be either perfectly equal or continually tending to equality. If in the same neighbourhood, there was any employment evidently either more or less advantageous than the rest, so many people would crowd into it in the one case, and so many would desert it in the other, that its advantages would soon return to the level of other employments. This at least would be the case in a society where things were left to follow their natural course,

> where there was perfect liberty, and where every man was perfectly free both to chuse what occupation he thought proper, and to change it as often as he thought proper. Every man's interest would prompt him to seek the advantageous, and to shun the disadvantageous employment (Smith, 1776, Book I, Chapter X).

Smith is describing the theory of compensating wage differentials that explain why jobs with higher educational requirements (a disadvantage, in his words) must be balanced with higher pay (an advantage).

Smith did not know about present value and IRR, but he would not have been surprised to hear that jobs requiring more training must pay more.

## References

Smith, A. (1776, originally published) *An Inquiry into the Nature and Causes of the Wealth of Nations*, www.econlib.org/library/Smith/smWN.html

Mincer, J. (1958) "Investment in Human Capital and Personal Income Distribution" *Journal of Political Economy*, pp. 281-302, www.jstor.org/stable/1827422

## Appendix

If you really want to know why $e^{slope} - 1$ is the exactly correct way to measure the percentage change from an earnings function's slope coefficient, then read on.

We start with a semi-log equation, $\ln Y = bX$, where we ignore the intercept to make the math easier to work with, but it would be the same with an intercept or other Xs.

We evaluate the equation at two points, $X_1$ and $X_2$ so we have $\ln Y_1 = bX_1$ and $\ln Y_2 = bX_2$.

We can take the difference, $\Delta \ln Y$, subtracting one from the other, $\ln Y_2 - \ln Y_1 = b(X_2 - X_1)$.

A property of logs is that subtraction is like division in own units so we can say this:

$$\Delta \ln Y = \ln(Y2) - \ln(Y1) = \ln(\frac{Y_2}{Y_1})$$

.

Now we can work on $\frac{Y_2}{Y_1}$ and do a clever little plus and minus move:

$$\frac{Y_2}{Y_1} = 1 + (Y_2 - Y_1)/Y_1$$

In case you cannot see it, I added a 1 and also subtracted a 1 since $\frac{Y_1}{Y_1} = 1$.

So, $\frac{Y_2}{Y_1} = 1 + \%\Delta Y$ and THIS IS WHERE THE 1 COMES FROM.

I am defining $\%\Delta Y$ as a decimal, say 0.5, and in percent it would be 50%.

We can substitute this expression for $\frac{Y_2}{Y_1}$ into $\Delta \ln Y = \ln(\frac{Y_2}{Y_1})$, which gives, $\Delta \ln Y = \ln(1 + \%\Delta Y)$.

When we take the antilog, the $e$ and *ln* cancel so that $e^{\Delta \ln Y} = 1 + \%\Delta Y$.

If we rearrange, we get $\%\Delta Y = e^{\Delta \ln Y} - 1$. HERE IS THE $-1$ AGAIN.

From the very first step we know that $\Delta \ln Y = b(X_2 - X_1)$. For a 1 unit change in X, $X_2 - X_1 = 1$, we get that $\Delta \ln Y = b$. Substituting that in to the $\%\Delta Y$ equation, we get:

$$\%\Delta Y = e^b - 1.$$

Ta da (aka QED).

For small b, we can say that $\%\Delta Y \approx \Delta \ln Y$ so $\%\Delta Y \approx b$. This is the approximation approach. The exact approach is $e^b - 1$.

The bigger b, the worse the approximation.

For example, we regress the natural log of house price (P) on the number of rooms in the house (R) and get $\ln P = 9.41 + 0.306R$.

We could use the approximation approach and say that one more room increases house price by about 30.6%, but that is a pretty big coefficient so the approximation might not do so well.

To get the percentage change in house price for a one room increase exactly right, we compute $e^{0.306} - 1$, which is 35.8%. In this case, the approximation does not work so well and we should use the exact approach.

# 8. Experience and Experience$^2$

We return to our earnings function, with the natural log of INCWAGE, and add another factor that affects earnings, experience. At first, additional experience increases earnings; but past a certain point, more experience lowers income. Thus, we will model experience with *two* additional *Xs* —experience and experience squared.

Do not be confused. Our goal is to measure the effect of education on earnings, but we are extremely concerned about omitted variable bias so we add experience to control for its effect on income. This does not fix omitted variable bias on the education slope coefficient because there are many other missing *Xs*, but it does improve our analysis.

## 8.1 Measuring and Modeling Experience

The ASEC survey (the March CPS that has extra questions about work and income) does not ask how many years of work experience the respondent has amassed. Our workaround is to assume that workers have been employed (and gaining experience on the job) throughout their adult lives whenever they were not in school. This is wrong, because adults can choose not to work, but it does give a rough measure of experience.

*STEP* Insert a column to the right of EducYears in your ASEC data sheet, select the cell in the top row, and label it *Experience.* In the cell below this label, enter a formula that computes Experience = AGE − EducYears − 6 and fill it down. We subtract 6 because most children start school at 6 years of age in the United States.

*STEP* Create a histogram of Experience. Make the bin width 1.

We have a problem because some of the values are negative and that does not make sense for this variable.

*STEP* Sort the data by Experience from low to high.

A handful of people 25 years old with 20 years of education have experience of −1 year. Our rough measure of experience failed for these people who skipped a grade or otherwise managed to get a post-graduate degree of some kind by 25 years of age.

*STEP* Manually change all negative values of Experience to 0.

We can draw a scatterplot and run a bivariate regression to explore the relationship between experience and income.

*STEP* Create a scatter chart. Make sure you have Experience on the $x$ axis. Regress INCWAGE on Experience via the Trendline and LINEST methods.

The chart is almost impossible to read and the fitted line is almost horizontal, with an intercept at roughly $80,000 and a really small slope coefficient. The interpretation is that one more year of experience is associated with about $6 more income (for 2024 ASEC data).

LINEST shows the same coefficients, of course, but also a large SSR and almost zero $R^2$. This is not good. What is going on?

The conditional mean function, CMF, which is also known as an experience-earnings profile, reveals the problem.

*STEP* Make a PivotTable of average INCWAGE given Experience. Copying an existing table and editing it is a time-saving approach, but creating it from the data sheet is also straightforward. Make a scatter plot connected by straight lines from the PivotTable.

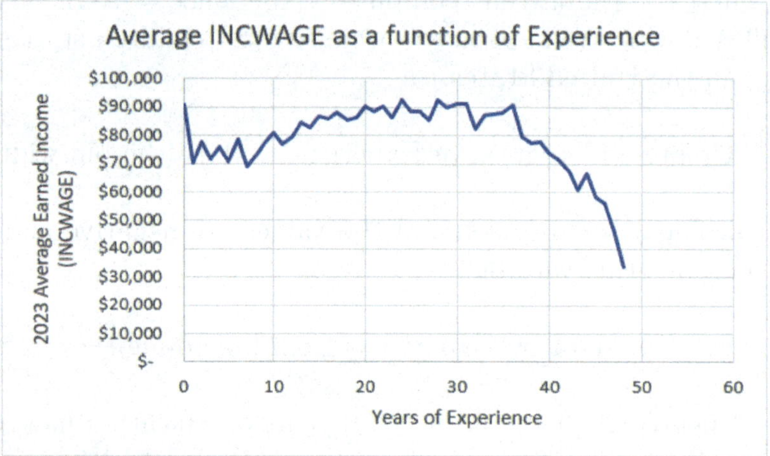


Figure 8.1: Income and experience have a curved relationship.

Figure 8.1 shows the result for 2023 income from the 2024 ASEC survey. The regression that we ran earlier is fitting an almost horizontal line to the non-linear relationship. For younger workers, more experience increases income, but this reaches a peak and then more experience lowers income. The line that minimizes the sum of squared residuals balances off the left and right parts of the curve, leaving us with a horizontal line that poorly describes the relationship.

Can linear regression handle this curvature in the data? Of course, it can. We just need the appropriate transformation of the variables and functional form. An easy solution, which is commonly applied, is a quadratic, $y = ax + bx^2$, with $a > 0, b < 0$.

*STEP* Add a sheet to the workbook. Name cells A1 and A2 "a_" and "b_" and enter values of 1 for both coefficients. Enter labels "a" and "b" in cells B1 and B2. Create a series for $x$ from 1 to 10 in cells C2:C11 with the label "x" in cell C1. Make $x^2$ in column D and create y in column E (applying the quadratic equation). Finally, make a scatter chart of $y$ as a function of $x$.

The graph does not look like it is going to generate a curve like the one needed for Figure 8.1. This is because the coefficients are both positive.

*STEP* Change $a$ to 25 and $b$ to $-2$.

This demonstrates that $a > 0, b < 0$ produces a hill shape. All we have to do is add an Experience Squared variable to our regression. The minimizing SSR procedure will choose the coefficients that best fit the data.

*STEP* Return to the data sheet and insert a column next to the Experience column. Label it "Experience Squared" and in the cell below it, enter a formula that squares the cell to its left.

Now we are ready to regress INCWAGE on Experience *and* Experience Squared.

*STEP* Select a range that is 5x3 because we have three coefficients (intercept, slope on Experience, and slope on Experience Squared). Enter the LINEST formula, making sure to select the $X$ range for both experience columns.

As expected, the coefficients are positive for Experience and negative for Experience Squared. But how is the fit?

STEP Return to the PivotTable sheet and create a Predicted INCWAGE column next to the Experience PivotTable. Use the LINEST coefficients in "backwards order" and square experience for the third coefficient. The formula ends up looking something like this, with Experience in column A in the PivotTable sheet and the regression coefficients in a sheet called rawdata: *=rawdata!$U$10+rawdata!$T$10*A4+rawdata!$S$10*A4^2.*

The moment of truth comes when Predicted INCWAGE is added to the CMF chart.

STEP As usual, the easiest method is to add data to a chart is to copy and edit the SERIES formula, but you can also click on the chart, click on the *Design* tab, and click *Select Data.*

Figure 8.2 shows the result. It makes clear that adding Experience Squared to the regression is an effective modeling strategy that captures the curved relationship between income and experience.

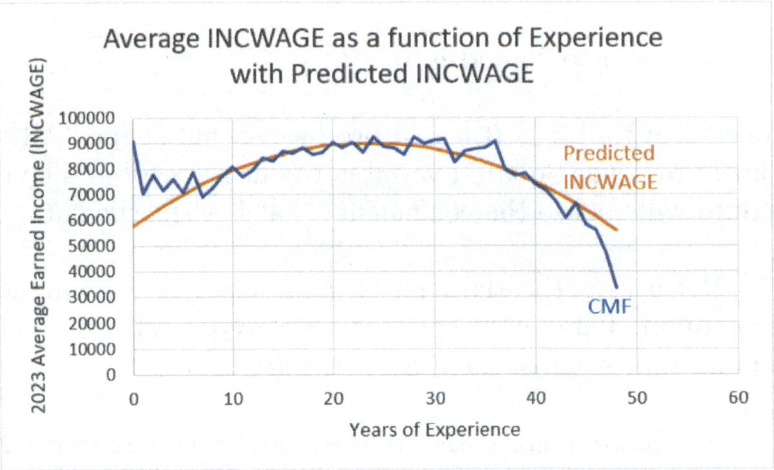


Figure 8.2: Adding Experience Squared gives a nicely fitted curve.

The fit is poor at very low and very high levels of experience, but there are fewer observations with these values of experience. The fit is quite good for 5 to 45 years of experience.

Importantly, when we adopt the *X+X^2* strategy to fit a curve, to evaluate the effect of a unit increase in $X$, you must use both coefficients and a specific value of $X$ since $\frac{dy}{dx} = a + 2bx$. Unlike the usual case where the slope tells you the rate of change from any value of $X$, including the squared term complicates the interpretation of the effect of $X$.

For example, at five years of experience, the regression for 2024 ASEC data tells us the rate of change in Predicted INCWAGE is $2711 - 2 \times 54 \times 5 \approx 2140$. An additional year of experience at this point increases income by roughly \$2,140. At 40 years of experience, however, we get $2711 - 2 \times 454 \times 40 = -1850$. We are on the downside of the experience-earnings profile and more experience lowers income.

As we will see in the next section, experience is not our primary interest. We include it as a statistical control to improve our understanding of the effect of education on income. Since experience has a non-linear relationship with income, we also include its square in the regression so that the $a > 0, b < 0$ coefficients capture that non-linearity.

## 8.2 Omitted Variable Bias in EducYears

Before we add experience to our regression of income on education, we explain what is going on. In previous work, we took the natural log of INCWAGE and regressed it on EducYears.

With this semi-log functional form, the interpretation of the slope coefficient is that one more year of education is associated with 12.4% higher income, on average.

Unfortunately, this is not a good measure of the effect of education because it is holding nothing else constant. We are ignoring variables correlated with education that also affect income and this produces omitted variable bias in the slope coefficient.

We do not explicitly say it, but we usually want to know something like, focusing just on education itself, how would one more year of education affect income? We want to *isolate* education and determine just its effect. Unless we include other variables that affect income in our regression, we do not get a good (unbiased) answer to our question.

There are many examples of raw comparisons with no controls that are misleading because of confounding (a synonym for omitted variable bias). Suppose someone claims a private high school is much better than the local public school. As evidence, they offer the fact that SAT scores at the private school are much higher than the public school.

The comparison of SAT scores can get quite sophisticated, with computations that show that the difference is statistically significant. This does not matter. The analysis is hopelessly flawed because nothing is being held constant.

It is easy to see that there are many factors affecting a student's SAT score and the school attended is just one of them. These factors produce school populations that are different for the private and public school. Elite Academy is going to do better than the public school because their rich students from stable households with parents that read to them when they were little have a huge head start. It could be true that the private school is better, but not as much better as the raw difference of SAT scores conveys.

The private school example was chosen because the confounding is obvious, but it is usually much harder to spot. WWII veterans had longer life spans than those who did not go to war. Would we conclude that war is good for your health? What is going on here?

*STEP* Watch Prof. Joshua Angrist explain this to puzzling outcome in this 3-minute video: tiny.cc/ww2bias.

Angrist uses the term *selection bias* which is a way to get omitted variable bias. His book, *Mostly Harmless Econometrics*, focuses on experimental design and how to isolate effects.

Omitted variable bias is the bane of everyone trying to make a causal statement or claim about a relationship between two variables. It produces an awkward situation where we almost always know our answer suffers from omitted variable bias, but holding constant at least some confounding variables is better than nothing.

*STEP* Regress ln INCWAGE on EducYears, Experience, and Experience Squared. The cell range for LINEST has to be 5x4 this time because we have four coefficients.

For 2024 ASEC data, we get:

$$\text{Predicted}lnINCWAGE = 8.9326 + 0.1226EducYears + 0.0277Experience \\ - 0.00045ExperienceSquared$$

Notice that the slope coefficient on EducYears has changed compared to the bivariate regression of ln INCWAGE on EducYears. Education is now associated with a roughly 13.5% increase in income, holding experience constant. Thus, education has a stronger effect on income than it did in the regression with no controls.

This "holding constant" language is critical. It tells us the other factors included in the regression. We do not mention both Experience and Experience Squared because they are not two separate *Xs*, we are modeling experience with these two variables.

To emphasize that we are holding experience constant and reinforce the interpretation of the slope as a percentage change in income, we can build a table that uses the regression equation to predict ln INCWAGE.

STEP Visit tiny.cc/interpreteduc. The table is shown in Figure 8.3, but online you can click on cell O2 to see its formula. Explore the other cells.

$\frac{\Delta\text{Pred ln INCWAGE}}{\Delta\text{Educ}}$ is the change in the natural log of income as education goes up by one. Predicted INCWAGE (the formula is *=EXP(ln INCWAGE)*) is rising at an increasing rate, which you can see in $\frac{\Delta\text{Pred INCWAGE}}{\Delta\text{Educ}}$ (notice it is not constant). However, the last column, the percentage change in INCWAGE for each additional year of education is constant and exactly equal to $e^{\text{slope of EducYears}} - 1$ in cell I8. This shows that our interpretation of the slope coefficient on EducYears in cell I8 is correct.

| Experience Squared | Experience | EducYears | Pred ln INCWAGE | ΔPred ln INCWAGE / ΔEduc | Pred INCWAGE | ΔPred INCWAGE / ΔEduc | %ΔINCWAGE / ΔEduc |
|---|---|---|---|---|---|---|---|
| 400 | 20 | 12 | 10.77644539 | | $47,880 | | |
| 400 | 20 | 13 | 10.89902052 | 0.1225751307 | $54,123 | $6,244 | 13.04% |
| 400 | 20 | 14 | 11.02159565 | 0.1225751307 | $61,181 | $7,058 | 13.04% |
| 400 | 20 | 15 | 11.14417078 | 0.1225751307 | $69,160 | $7,978 | 13.04% |
| 400 | 20 | 16 | 11.26674591 | 0.1225751307 | $78,178 | $9,019 | 13.04% |

Figure 8.3: Interpreting the slope coefficient on EducYears.

STEP Change Experience to 10 and then 30.

This has no effect on the percentage change in INCWAGE. This shows that we can interpret the slope coefficient on education as having the same value given any experience value. The key is not a particular numerical value of experience, but that it is *held constant* as education changes.

Have we corrected the problem so we have a bias-free measure of the effect of education on income? Of course not. We only included one factor as a statistical control. That is a long way from the unreachable ideal of ceteris paribus, *everything else* held constant.

Surely income is affected by ability and motivation. Earlier, we called this Drive. To the extent that these factors are correlated with education, then omitting them from the regression increases the slope coefficient on education. It makes it seem like education is responsible for more of the increase in income than it deserves. Education is getting "too much credit" because it is implicitly including the effects of ability and motivation since they are not included in the regression.

Not every omitted variable will decrease the slope coefficient on EducYears. Experience, for example, did not. It slightly raised it. That has to do with the relationship between EducYears and experience and how experience affects income. Sometimes, an omitted variable has little effect on the coefficient of an included $X$. That will happen when the omitted and included Xs have only a slight degree of connection or if the omitted variable has a weak effect on the dependent variable.

## 8.3 Judging the Earnings Function

There are countless papers that measure the rate of return on education as the coefficient on years of schooling in a regression with a dependent variable that is the natural log of some monetary amount (such as earnings, income, or wages), like this:

$$Predictedln\$ = b_0 + b_1 EducYears + b_2 Experience + b_3 ExperienceSquared$$

You learned earlier that this functional form is strongly supported by economic theory because dollars earned in the future must be present valued. You also know that by adding more statistical controls (more independent variables, such as demographic characteristics or measures of drive and intelligence), we can improve the estimate of the coefficient of education.

However, the earnings function, while often used, is certainly not perfect. Along with fears of omitted variable bias, there are other serious issues with interpreting the coefficient on schooling as a rate of return to education in an earnings function regression.

First, we have no information on out-of-pocket costs of education. Second, we are using past returns and assuming they will repeat themselves in the future. This is a heroic and unreasonable assumption. This would be like estimating the expected return of the stock market using historical returns —everyone knows that "past performance is not indicative of future results" and leads to the funny quip, "That's like driving a car by looking at the rearview mirror."

We simply do not have the required information to calculate a true rate of return, as used in the financial world. We do not have tuition and other costs. More importantly, we do not know future returns for different years of schooling obtained today. But, these clear weaknesses notwithstanding, earnings functions with many different combinations of statistical controls continue to be estimated because they do give us a view, albeit imperfect, of the effect of education on economic outcomes.

## 8.4 Takeaways

Previously, we used the natural log of INCWAGE to get a curved relationship between INCWAGE and EducYears. In this section, we used the CMF to see that experience affects income in a non-linear way.

Earlier, we saw how a scatterplot of INCWAGE on EducYears is just a bunch of dots, but the CMF reveals the pattern. In the same way, the average INCWAGE as a function of experience reveals a clear curved shape which we can model by a squared term.

Regression is incredibly flexible. By using Experience and Experience Squared together, we matched the curved effect of experience on income.

We want our model to be properly specified so that it does not suffer from omitted variable bias. Adding experience improves our estimate of the effect of education on earnings, but it does not make it perfect. We still have omitted variable bias.

One of the biggest weaknesses of observational studies is that individuals choose what to do. The ASEC survey simply asks people how much education they have and their income. There are many differences between people with a college degree and those with a high school degree other than the educational difference. We have to find these other factors and include them in the regression to get a better measure of the effect of education on income, holding other factors constant.

While we can certainly improve our estimate of the effect of education, we will never have a true, ideal rate of return measure because we do not have costs or forward-looking, future returns. A perfect estimate of the rate of return to education is impossible.

## References

Angrist, J. and Pischke, J-S. (2009) *Mostly Harmless Econometrics*, Princeton University Press, www.mostlyharmlesseconometrics.com.

# 9. Dummy Variables

Not only can regression handle non-linear relationships between $Y$ and $X$ through appropriate variable transformations, it can also easily incorporate qualitative or categorical data with *dummy variables*, also known as binary or dichotomous variables because they have only two values: 0 or 1.

Our primary goal is to learn how to interpret the coefficients on dummy variables. This can be tricky, so we will proceed slowly.

We continue with our earnings function and 2024 ASEC data from IPUMS CPS. By adding more information to our regression, we get an even better read on the effect of education on income.

## 9.1 Male and Female

The Census Bureau and Bureau of Labor Statistics do not provide people the opportunity to identify as non-binary. Perhaps this will change in the future, but for now, people are asked their sex and they must choose between male and female.

Before we make a dummy variable called Male, we need to know the codes. We can click on the SEX variable in IPUMS CPS to see that the codes are 1 for male and 2 for female.

You may continue working with your data for the most recent year or go to tiny.cc/interpretdummy to work with 2024 ASEC data. Using the latter allows you to exactly replicate the numbers in this section.

*STEP* Insert a column (if needed) after Experience Squared (from previous work) and label it *Male* in row 1. In the cell below enter the formula *=IF(B2=1,1,0)*. This assumes the SEX variable is in column B. If not, use the column for the SEX variable. Fill it down.

You can see that whenever SEX is 1 in column B, it is 1 for Male; while the codes of 2 have been recoded to zero—meaning that the person is not a male.

*STEP* Insert a column (if needed) after the Make column and label it *Female*. Enter an IF statement that produces 1 if the person is a female and 0 otherwise. Fill it down.

It is easy to see that whenever Male is 0, Female is 1 and vice versa. They are opposites. Notice also that they sum to one for every observation.

*STEP* Compute the average of the Male and Female dummy variables. What do you notice?

The answer is that the averages sum to exactly one. Less obvious is that the average of a dummy variable tells us the percentage of observations that has the characteristic. Thus, there are about 53% men and 47% women in our 2024 data.

Most people know that men earn higher incomes than women, but how much more?

*STEP* Create a PivotTable that shows the average INCWAGE for men and women.

In the 2024 ASEC for our group of 25-64 year-old people that made at least $20,000 in earned income, men make about $90,000 and women earned a little over $70,000, on average. That almost 25% difference is a lot and is the subject of much research.

But we do not need a PivotTable to compare average incomes of men and women. Regression gives us the exact same information.

*STEP* Create a chart of INCWAGE as a function of Male and fit a trendline, displaying the equation. Be careful. Male needs to be on the $x$ axis and there are two vertical lines with dots (at $x$ axis values of 0 and 1).

For 2024 ASEC data, the fitted regression line is:

$$PredictedINCWAGE = 71227 + 21814 Male$$

The intercept tells us how much income women earn, on average, and the slope coefficient tells us how much more than women, on average, men make. It is easy to see that this exactly agrees with the PivotTable results.

To see how the dummy variable's coefficient works, we compute Predicted INCWAGE for women and men. For women, we substitute in a 0 for Male, leaving only the intercept. This is called the *base case*. For men, Predicted INCWAGE is the intercept plus the slope (since Male = 1).

What happens if we use the Female dummy variable?

*STEP* Create a chart of INCWAGE as a function of Female and fit a trendline, displaying the equation.

The result clearly shows that the slope is the same as before, but of opposite sign. The intercept is now the average income for men and men are the base case. We add the slope coefficient on Female (which is a negative number) to find women's average INCWAGE.

The base case depends on how you construct the dummy variable. If we had a dummy variable called Married, the base case would be not married. The base case is always the category that has a dummy variable value of zero.

What happens if we include both dummies at the same time?

*STEP* Use LINEST to regress INCWAGE on Male and Female.

Excel's LINEST and most statistical software drops one of the variables. Google sheets shows an error. The way dummy variables work is that there always has to be a base case. We cannot fit a line using both Male and Female so some routines simply arbitrarily drop of one of the categories, while others report that coefficients for the regression equation cannot be found.

The lesson is that we never include all of the categories as dummy variables in a regression. If we had three colors, Red, Blue, and Green, we would have three dummy variables, but we include only two of them in the regression. The one not included would be the base case.

## Adding Male to the Earnings Function

We return to our earnings function regression and measuring the effect of schooling on income. In previous work, we saw that a semi-log functional form is the correct specification and that a quadratic modeling of experience captures the non-linear relationship between experience and income.

Now, we extend this analysis by including information on whether the observation is male or female in the earnings function. Our research question remains focused on the effect of education on income, but by including the variable SEX, recoded as a Male dummy variable, as a statistical control, we get a better measure of how education affects income.

*STEP* Use LINEST to regress ln INCWAGE on EducYears, Experience, Experience Squared, and Male.

Figure 9.1 shows the result for 2024 ASEC data. The slope coefficient on EducYears has slightly risen again (compared to the regression without Male). The interpretation is that one more year of schooling, holding constant experience and sex, is associated with about a 13.85% ($e^{0.12975} - 1$) increase in INCWAGE.

| Male | Experience Sq | Experience | EducYears | intercept |
|---|---|---|---|---|
| 0.3026596224 | 0.000432116908! | 0.02691383869 | 0.1297500996 | 8.670292826 |
| 0.005042439561 | .0000202235072 | ).00098062057 4 | 0.001045683788 | 0.01970526577 |
| 0.267364817 | 0.5592996207 | #N/A | #N/A | #N/A |
| 4572.554519 | 50119 | #N/A | #N/A | #N/A |
| 5721.474059 | 15678.0284 | #N/A | #N/A | #N/A |

Figure 9.1: Regressing ln INCWAGE on EducYears and two controls.

We also can say that men earn, on average, about 35% more than women ($e^{0.303} - 1$), holding constant education and experience. Notice how the coefficient is relatively far from zero so the approximation (just directly reading the coefficient) is off by 5 percentage points.

We can also make a chart that will help us understand that dummy variables are intercept shifters.

*STEP* Create a vertical range of cells for EducYears from 10 to 20 by one. Create formulas for Predicted ln INCWAGE for Male and Female with 20 years of Experience using the multiple regression coefficients (in Figure 9.1). Make a scatter chart with straight lines.

Figure 9.2 shows the result for ASEC 2024. The two lines are parallel and the vertical difference is the slope coefficient on Male. In a regression, slope coefficients on dummy variables are seen as *differential intercept shifters.* They are interpreted by adding them to intercept for the group that has the characteristic.

| EducYears | Male | Female |
|---|---|---|
| 10 | 10.63588345 | 10.33322383 |
| 11 | 10.76563355 | 10.46297393 |
| 12 | 10.89538365 | 10.59272403 |
| 13 | 11.02513375 | 10.72247413 |
| 14 | 11.15488385 | 10.85222423 |
| 15 | 11.28463395 | 10.98197433 |
| 16 | 11.41438405 | 11.11172443 |
| 17 | 11.54413415 | 11.24147453 |
| 18 | 11.67388425 | 11.37122463 |
| 19 | 11.80363435 | 11.50097473 |
| 20 | 11.93338445 | 11.63072483 |

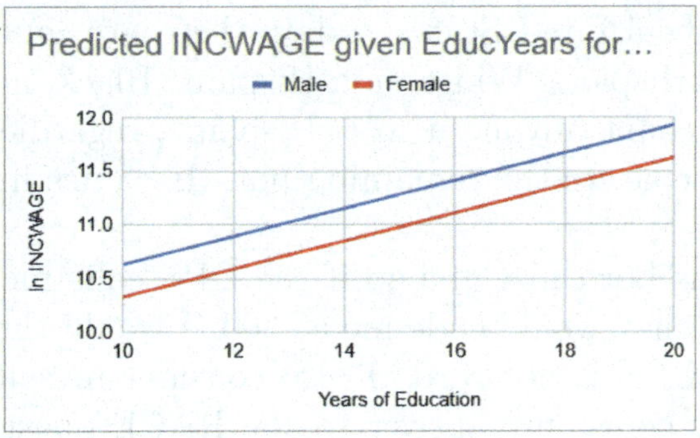


Figure 9.2: The Male coefficient shifts the intercept.

Notice that the slope in Figure 9.2, the change in ln INCWAGE as education increases, is the same for men and women. There is something called an interaction term that enables relaxing this restriction, but this is beyond the scope of this book.

## 9.2 Race and Ethnicity

As a second example of how dummy variables work as differential intercept shifters, we work on how race and ethnicity can be incorporated into a regression equation.

As before, you may continue working with your data for the most recent year, but you will have to return to the IPUMS CPS and revise your extract, adding RACE and HISPAN. You then recreate EducYears, Experience (remember to fix the few observations with negative experience), Experience Squared, and Male. Or, you can go to tiny.cc/interpretrace to work with 2024 ASEC data that has been prepared for you.

The CPS asks separate questions about race and ethnicity. The codebook shows many categories for the variable RACE, but we will focus only on those who identified as White (100), Black (200), and Asian (651). The variable HISPAN is 0 for non-Hispanics.

This is confusing. Let's be clear. All of these categories are socially constructed and have no basis in substantive biological differences. White, Black, Asian, Pacific Islander, and American Indian rely on differing physical appearance and are called races. Ethnicity is a different thing. It is based on cultural differences. Hispanics can be any race.

In the United States today, there are four major demographic categories: non-Hispanic White, non-Hispanic Black, non-Hispanic Asian, and Hispanic. We count anyone who is Hispanic, regardless of race, as Hispanic and then we separate the remaining non-Hispanics into the three racial groups.

I was born in Cuba so if the CPS calls me, I would tell them I am Cuban and they would code me as 300. I would then be Hispanic or Latino because HISPAN is not zero. I also consider myself White so I would be 100 when they coded my answer to the RACE question. Even though I say I am in the White race, I am not a non-Hispanic White. Going forward, when we say White, Black, or Asian, we mean those races, but only for people who do not have Hispanic ethnicity.

*STEP* Proceed to the *Pivot Table 1* sheet in the Google sheet to see the number of observations in each category.

There are roughly 30,000 Whites, 5,000 Blacks, 4,000 Asians, and 10,000 Hispanics. The remaining observations will be grouped into an Other category. Also, you can see how Hispanics really are distributed across the three most populous races.

*STEP* Return to the *rawdata* sheet and enter the labels *White*, *Black*, *Asian*, *Hispanic*, and *Other*, in cells M1 to Q1. In cell M2, enter the formula =IF(AND(D2=0,C2=100),1,0) and fill it down. In cell M50126, compute the sum of the cells above.

The sum of the White dummy variable must be equal to 29,470 (cell I3 in the *Pivot Table 1* sheet).

*STEP* Create the other four dummy variables. Black and Asian are similar to White, but you do have to change the C2 cell to 200 and 651, respectively. Hispanic is simply 1 if HISPAN > 0. You can use this formula for Other: *=IF(AND(C2<>100,C2<>200,C2<>651,D2=0),1,0)*. Compute the sum of each dummy (like we did for White) and confirm that it equals the count value in the *Pivot Table 1* sheet. If it does not, something is wrong.

We include the first four dummy variables to our regression, but not Other. Why leave one out? Because we need a base case. It represents not being any of the included dummies. If you include all cases, the regression will crash like when we include Male and Female.

*STEP* Run LINEST with ln INCWAGE as the dependent variable and include EducYears, Experience, Experience Squared, Male, White, Black, Asian, and Hispanic.

The slope coefficient on EducYears is our primary focus. Adding ethnicity and race has lowered the return to education a little, but it is still hovering around 13% which is an excellent investment.

Adding the other *Xs* enables us to hold these factors constant as we vary education and helps us to isolate the relationship between education and income. We have not, however, come close to completely explaining the variation in income—$R^2$ has been rising as we have added Xs, but its value of almost 30% means we still have 70% of the variation in income unaccounted for.

The coefficients on the dummy variables can be interpreted by referring to the base case, Other. Since we have ln INCWAGE as the dependent variable, the dummy coefficients are an approximation to the percentage difference in income between the base case and having the characteristic.

Thus, in the 2024 ASEC, the coefficient on Hispanic was −0.046. This means, holding constant education, experience, and sex, Hispanics earned roughly 4.6% less than Other, which is the base case. You can read the coefficients on White, Black, and Asian the same way. You can also use the anti-log minus 1 to compute an exact percentage difference.

Does this mean Asians have higher incomes than Whites since their coefficient is higher? Yes, it does. If you do not believe this, you can make a Pivot Table to confirm it.

In fact, you can subtract the coefficients, Asian - White, and conclude that, holding constant education, experience, and sex, Asians have about 3.6% higher incomes than Whites. On the other hand, Blacks and Hispanics have much lower incomes than Whites (and Asians), holding constant education, experience, and sex.

These demographic income gaps point to discrimination, but we need to worry about omitted variable bias. There are many other factors not included in our regression that are confounding our coefficient estimates.

## 9.3 Takeaways

Dummy variables are a modeling strategy that enables regression to incorporate qualitative information and categorical variables. It is wrong to believe regression is limited to straight lines. In fact, with variable transformations, dummy variables, and other techniques, regression is incredibly flexible.

Dummy variables should have the name of the characteristic if true. Thus, Male tells us that 1 means male and 0 does not. Sex is a poor name because it does not tell us what 1 and 0 mean.

A dummy coefficient is interpreted by comparing it to the base case. It conveys the difference between having the characteristic and not having it.

With a semi-log earnings function, the coefficient on a dummy variable is the percentage difference between having and not having the characteristic. As with continuous variables, the coefficient is an approximation to the percentage change. It can be exactly computed by $e^{coefficient} - 1$.

> *EXCEL TIP* There is a dummy variable for each category. Suppose the variable *Transportation* had car, bus, and bike options. Use an IF statement to make the *Car* dummy variable equal to 1 if $Transportation = car$. Same for *Bus* and *Bike*.

With many categories connected to a single variable, one dummy variable coefficient can be subtracted from another to reveal the difference between those two categories.

If you are ever in doubt about the meaning of a coefficient, create scenarios that predict $Y$ and then compare the outcomes.

We have merely scratched the surface of modeling with dummy variables. An important modeling technique involves interaction terms. Multiplying a continuous variable by a dummy variable (such as EducYears × Female) and adding it to the regression enables the slope coefficients of the groups to vary.

We almost never create dummy variables by hand, like we did here. Modern statistical software, unlike a spreadsheet, can easily create dummy variables for us. But it is good to know how they work and how to interpret the coefficients—do not forget the base case!

# 10. Monte Carlo Simulation

Monte Carlo simulation simply means the repeated running of a chance process and then direct examination of the results. We will use it as a foundation for interpreting regression coefficients as random variables. The estimated coefficients come with associated standard errors which are produced by regressions from data that contain randomness. The highest priority is understanding the *sampling distribution* of a statistic.

Monte Carlo simulation is powerful for research, but also for teaching and learning. Its core is brute force repetition and then examination of the results. It requires no imagination at all. It will be our go to method for understanding randomness and the sampling distribution.

## 10.1 Gauss and Two Approaches

Carl Friedrich Gauss (1777-1855) was mentioned in the preface to this book as the GOAT in mathematics. Before the euro, the ten-Deutsch mark note featured him along with a graph of the normal curve (which he made famous and is called the Gaussian distribution). Look carefully in Figure 10.1 and you can see that it even displays the equation of the normal distribution.

Figure 10.1: German currency featuring Gauss.

There is a story, probably apocryphal, of how he amazed his kindergarten teacher. Apparently, the children were especially unruly one day, so the teacher punished them. He told them to add all the numbers from 1 to 1000. This starts easily, but gets tedious and painful pretty quickly. $1 + 2 = 3$, $3+3 = 6$, $4+6 = 10$, and $5+10 = 15$. It will take a long time to get to 1,000.

After a few minutes, Gauss raised his hand and announced the answer, 500,500. The teacher asked him where he got that number, which is correct, and Gauss said he noticed a pattern. Remember, he is five years old.

If you make a list of the numbers, then create a second list, but flipped, the pairs always add up to 1,001. 1 goes with 1,000, 2 with 999, 3 with 998, and so on until the end when 998 is with 3, 999 with 2, and 1 with 1,000.

The rest is easy (well, maybe not for the usual five-year old, but this is Gauss). Multiply 1,001 by 1,000 (since there are 1,000 pairs) and divide by 2 to get 500,500. As they say, QED.

This is clever, remarkable, and beautiful. It is like Michelangelo and the Sistine Chapel. Is Monte Carlo simulation like this? No.

Monte Carlo simulation is a different approach to problems which uses little creativity or subtlety. It is a direct attack on a question.

The analogous way of solving the teacher's tedious problem with simulation is to use a spreadsheet to add the numbers.

*STEP* Make a list from 1 to 1000 in cells A1:A1000 and then, in cell B1, enter the formula *=SUM(A1:A1000).*

Excel displays 500,500. This is nowhere as magnificent as what Gauss did, but it does give the answer.

Monte Carlo simulation was developed during World War II by physicists working on the Manhattan Project. Nicholas Metropolis coined the term because his colleague, Stanislaw Ulam, was an avid poker player. They were simulating how radiation propagates and incorporating randomness. The connection to chance and gambling is why Metropolis named the method after the famous casino in Monaco.

So Monte Carlo simulation, or simulation for short, is an alternative to the analytical approach. Instead of using formulas and derived results, simulation uses computers to repeat the chance process many times and then directly observes the outcomes.

This is too abstract an explanation. We create a concrete example to show how it works.

## 10.2 Are You Really a 90% Shooter?

You walk into the gym and announce that you are a 90% free throw shooter. This is impressive. Someone hands you a basketball and says, "Prove it."

You shoot 100 free throws, but it does not go as well as you hoped. You make only 75.

The question is: Are you really a 90% shooter? After all, you could be and you just had some bad luck. By chance, you missed a few more than expected. So, another way to ask the question is: Can randomness explain this poor outcome?

Our solution strategy will be simulation, but make no mistake, Gauss would not have needed simulation. He would have immediately rejected your claim. He knows a formula that can be used, $\sqrt{n}\,\sigma$, which answers the question quickly. The formula is the product of sophisticated mathematics and can be called beautiful, but most people find it extremely difficult to understand and cannot use it to answer the question.

We begin with RAND(). As a reminder, Excel draws uniformly distributed random numbers in the interval from 0 to 1 with the RAND() function.

*STEP* In cell A1, enter the formula *=RAND()*. Repeatedly press F9. This is the keyboard shortcut to recalculate the sheet.

To model a 90% free throw shooter, we use an IF statement.

*STEP* In cell B1, enter the formula *=IF(A1<0.9, 1, 0)*.

When the IF function evaluates to 1, the free throw is made and when it is 0, it is missed. This is a binomial random variable since it can only take on two values, 0 or 1.

We do not need to actually see the random number generated so we can embed RAND() directly in the IF statement.

*STEP* In cell B2, enter the formula *=IF(RAND()<0.9,1,0)*. Fill down this formula to cell B100.

This is the *data generation process* or DGP. It tells us how our data are produced.

STEP Rename the sheet (double-click on the sheet tab) *DGP*.

As you scroll back up to the top row, you will see many ones and a few zeroes. With a 90% success rate, roughly one in ten cells will have random number greater than than 0.9 and, therefore, show a 0.

The fact that each cell in column B stands alone and does not depend on or influence other cells means we are assuming *independence*. In our model, a miss or make does not affect the chances of hitting the next shot.

If you believe in the hot hand (Cohen, 2020), this implementation of the chance process is wrong. If making the previous shot increases the chances of making the current shot, there is *autocorrelation* and we cannot use 0.9 as the threshold value for every shot.

How many shots out of 100 will a 90% shooter make?

STEP Enter the formula *=SUM(B1:B100)* in cell C1 and the label, "Number made" in cell D1.

You see a number around 90 in cell C1. Each press of F9 gives you a new outcome from 100 free throws. The number of made free throws is not always exactly 90 because there is chance involved in making a shot.

STEP Press F9 to recalculate the sheet a few times to get sense of the variability in the number of made shots from 100 free throws.

There is no doubt that the number of made shots is a random variable since it is bouncing around when you recalculate the sheet. It makes common sense that adding binomial random variables will produce a random outcome.

A *statistic* is a recipe for the data. Cell C1 is a statistic because the recipe is to add up 100 shots. We are interested in the distribution of the sum of 100 free throws from a 90% free throw shooter, including its central tendency, dispersion, and shape of a histogram of outcomes. With this, we can decide if 75 made shots is merely unlikely or so rare that we reject the claim that you are a 90% shooter.

Simulation is simply repeating the experiment many times so we can approximate the center, dispersion, and distribution of the outcomes. Since we are working with a statistic, the distribution of the sum of 100 free throws is called a *sampling distribution*.

To process the many outcomes, we need software. A free Excel add-in that does Monte Carlo simulation is available here: dub.sh/regcourse.

*STEP* We used it earlier for the BIVARNORMAL function, but if needed, download the *MCSim.xla* file from the link above and use the Add-Ins Manager (File → Options → Add-ins → Go) to install it. Click the *Add-ins* tab and click *MCSim.*

Figure 10.2 shows the MCSim add-in dialog box. On the left are three required choices. You must select a cell to track (C1 in our example), the number of repetitions (the default is 1,000), and the random number generator (RNG) to use. The MCSim add-in comes with its own RNG, RANDOM. Selecting it will replace all RAND in the sheet with RANDOM. The default is no changes.

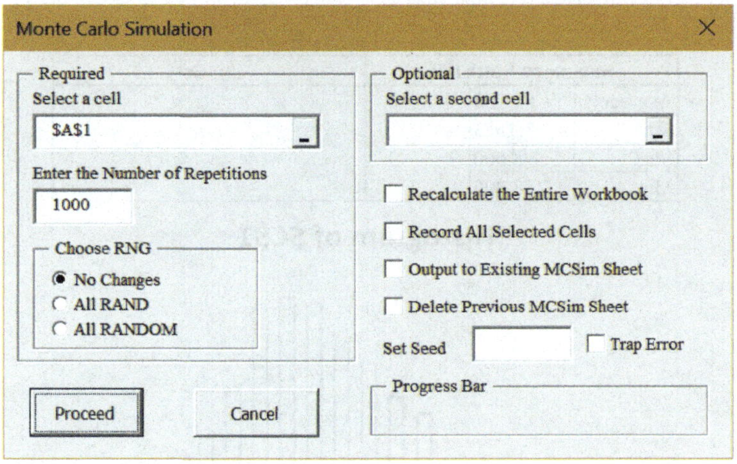


Figure 10.2: The Monte Carlo Simulation Excel Add-in.

On the right are some advanced options. Some of these will be used in future work. The Set Seed option forces the RNG to begin from the same initial position, which allows for replication of results.

STEP Click in the *Select cell* box, clear it, and click cell A1. Click in the *Set Seed* box and enter 123. Click *Proceed.*

The progress bar indicates time remaining for the simulation and a new sheet is inserted in the workbook. It shows the first 100 outcomes in column B, summary statistics, and a histogram.

STEP Return to the *DGP* sheet and repeat the simulation of cell A1.

The results are exactly the same as before because the Set Seed option started the RNG from the same initial value.

STEP Return to the *DGP* sheet and click the *MCSim* button in the *Add-ins* tab. Select cell C1 and clear the *Set Seed* box. Click *Proceed.*

Figure 10.3 shows the results. Yours will be different because we cleared the *Set Seed* box. However, your results will be quite close in the sense that your average is near 90 and the SD is around 3. The average and SD are approximations to their true exact analogues, the expected value (EV) is 90 and the standard error (SE) is 3.

| Summary Statistics | | Notes |
|---|---|---|
| Average | 89.955 | |
| SD | 3.1523 | |
| Max | 98.000 | |
| Min | 78.000 | |

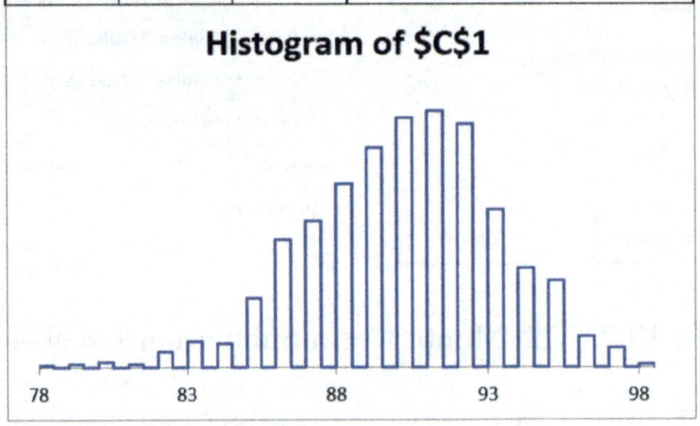


Figure 10.3: Simulation Results for Sum of 100 Free Throws from a 90% Shooter.

The sim also shows the maximum and minimum shots made from 100 free throws in 1,000 repetitions. In Figure 10.3, the max is 98 and min is 78.

This gives an answer to our question. With 1,000 sets of 100 attempts the worst a 90% shooter did was 78. So, we do not believe that you are a 90% free throw shooter.

What if we did more repetitions?

*STEP* Run a simulation of 10,000 sets of 100 free throws.

Once again, you are unlikely to see 75 or fewer. It seems the bad luck defense is not going to work. While it is possible that you really are a 90% free throw shooter and had an incredibly unlikely run of bad luck, such an outcome is incredibly rare. So rare, in fact, that we do not believe you.

Notice the average and SD values changed with the second simulation. Simulation always gives an approximation to the exact answer.

By the way, Gauss would have computed the SE with the formula $\sqrt{n}\,\sigma = \sqrt{100} \times 0.3 = 3$ and announced that $\dfrac{75-90}{3} = -5$. In other words, 75 is 5 standard units away from what we expected and this is ridiculously unlikely so we do not believe that you are a 90% free throw shooter.

## 10.3 Max Streak

A second example of simulation involves streaks, also known as runs. A streak in this case is a consecutive set of made shots.

*STEP* Return to the *DGP* sheet. Starting from cell B1, find the first 1 (it could be cell B1), and then count how many 1s in a row you see before you encounter a miss. Write that number down and compute the next streak. Continue until you reach the last shot attempt. The longest streak is the *max streak*.

The question is: what is the length of the typical max streak in a set of 100 free throws from a 90% free throw shooter?

This is an exceedingly difficult question. With the expected number of made free throws, it was easy to see that a 90% free throw shooter will make around 90 out of 100 attempts. But the max streak is a much more complicated statistic. How can we answer this question?

The analytical approach is a dead end. There are formulas that approximate a solution (Feller 1968, p. 325), but they are quite complicated. No exact analytical solution has been found.

Simulation can be used if we can figure out a way to implement the problem in Excel. This is the one area when simulation requires some ingenuity. After all, we need a cell that computes the max streak so we can use the MCSim add-in.

*STEP* In cell E1, enter the formula *=B1*. In cell E2, enter the formula *=IF(B2=1,E1+1,0)*. Fill it down a few cells.

Now you can see what the formula is doing. If the shot is made, we add it to the previous running sum, but if it is missed, it resets the running sum to 0.

*STEP* Fill the formula down to E100 and look at the values as you scroll back up.

You should see several streaks in a set of 100 free throws. We want the longest streak. That is easy.

*STEP* In cell F1, enter the formula *=MAX(E1:E100)* and enter the label "max streak" in cell G1. Press F9 a few times.

Cell F1 is displaying the max streak from each set of 100 free throws. Max streak is a statistic, just like the sum, because it is a recipe—albeit much more complicated than the sum.

Max streak has an expected value, standard error, and sampling distribution. We can approximate all of these with simulation.

*STEP* Run a simulation, with 10,000 repetitions, of cell F1.

Figure 10.4 shows results. Yours will be a little different. The average is an approximate answer to our question: the max streak is about 27 or so. The exact answer is the expected value, but we have no way of computing it.

| Summary Statistics | | Notes |
|---|---|---|
| Average | 26.858 | |
| SD | 9.5715 | |
| Max | 85.000 | |
| Min | 9.000 | |

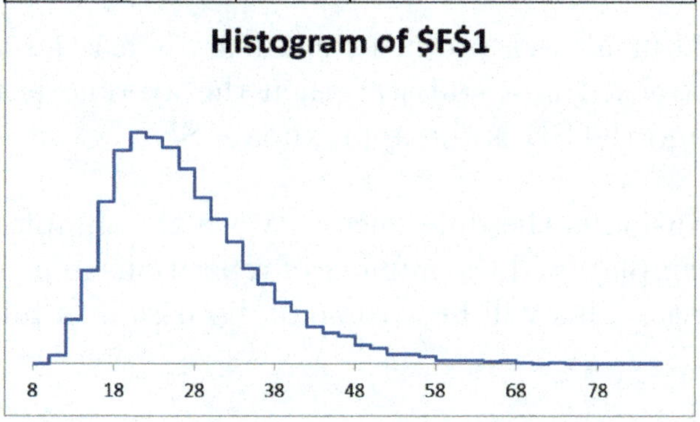


Figure 10.4: Sim Results for a 90% Shooter's Max Streak.

Of utmost importance is that the histogram is an approximation to the exact sampling distribution (which no one has figured out how to exactly derive). The graph tells us which values are unlikely: roughly 10 or fewer and 50 or more.

Notice that the sampling distribution of the max streak statistic, unlike the sum, does not appear to follow the normal curve. The max streak has a long right tail and is not symmetric.

## 10.4 Takeaways

Monte Carlo simulation is a brute force approach to answering questions involving chance. A much older alternative, the analytical approach, relies on brainpower to derive formulas.

To run a simulation, the problem must be implemented in Excel (or some software that can generate random numbers). Of course, one can manually flip a coin many times in the real world, but this is tedious. Simulation did not become a powerful tool until modern computers were invented and enabled a great many repetitions in a short period of time.

Simulation is always only an approximation. By running more repetitions, the approximation improves, but it can never give an exact answer because it would have to run forever.

Often, we are searching for the sampling distribution of a statistic. This tells us the chances of each outcome, the typical result (called the expected value), and the dispersion in values (called the standard error).

The MCSim add-in always produces summary statistics and a chart. If the cell that is tracked is a statistic, then the average is the approximate expected value and the SD is the approximate SE.

One area of confusion is the difference between the sample size, $n$ (100 free throws in our example), and the number of repetitions in a simulation. These are different things. This will be a topic of discussion in future work.

## References

Cohen, B. (2020) *The Hot Hand: The Mystery and Science of Streaks*, Custom House. Russ Roberts interviews Cohen in an August 10, 2020 episode of EconTalk available at www.econtalk.org/ben-cohen-on-the-hot-hand.

Feller, W. (1968, 3rd ed.) *An Introduction to Probability Theory and Its Applications*, John Wiley & Sons.

# 11. Polling DGP

We will study the properties of a data generation process (DGP) common in politics. There is a population of voters, some prefer candidate A and others B. We sample, that is, we take a poll, and use the results to estimate the shares for each candidate.

The Polling DGP is easy to understand and implement. We need a population, from which we randomly draw a sample, and then use a sample statistic to estimate its population analogue. A *statistic* is a recipe that tells you how to process numbers.

Our primary goal is to understand that the sampling distribution has an expected value (EV) and standard error (SE). The confusing part is that there are three SEs:

1. Exact SE
2. Approximate SE
3. Estimated SE

We walk through an example with fake data so that we can focus sharply on how sampling works, then return to the CPS ASEC survey and apply what we learned to a real-world question.

## 11.1 A Hypothetical Example

The Polling DGP assumes a fixed population. Each person's information is on a piece of paper in a box and we draw out a sample and read the paper.

For this hypothetical example, we will know the contents of the box, but usually the population is so huge that a census is too expensive.

*STEP* Create a series from 1 to 1000 by 1 in cells A1:A1000. In cell B1 compute the average and enter the label, "$\mu$, population average" in cell C1.

The population average, mu ($\mu$), is 500.5. This is a *parameter*—a number that is usually unknown about a population. It is common to use Greek letters for population-based parameters and their English-language counterparts for the same statistic computed with sample data.

*STEP* In cell B2 compute the SD using Excel's STDEV.P function and enter the label, "$\sigma$, population SD" in cell C2.

The population SD, sigma, is 288.675. These numbers are quite spread out. A histogram confirms this.

*STEP* Create a histogram of the population. Change the bin widths to 1 to see that we have one of each number.

With our population in Excel, we are ready to sample. We will use a function in the MCSim add-in called SAMPLE. The MCSim add-in must be installed for the function to work.

SAMPLE is an array function (like LINEST and BIVARNORMAL) with four arguments: SAMPLE(pop range, prob range, n, optional 1 if with replacement). Prob range is the probability of selection for each individual in the population.

A *simple random sample* (SRS) means that each individual is equally likely to be chosen. Most large surveys, like the CPS, are not SRS. They have what are called complex survey designs, with strata and clusters. In addition, the CPS oversamples Hispanics to increase the reliability of estimates for them. The CPS is not SRS because the probability of being selected is not the same for each person.

*STEP* Provide the prob range information in column J by entering the value 0.001 in J1 and filling it down to J1000.

This means each individual in the population has an equal, 1 in a 1,000, probability of selection. We are ready to draw a sample.

There are three steps:

1. Select the appropriate size range: the number of rows is the sample size (n) and the number of columns is 1 (for prob range) + the number of variables in the population.

2. In the formula bar, enter *=SAMPLE(*, select the appropriate cell ranges and other two arguments, then add the closing parentheses.

3. Press ctrl-shift-enter (the keyboard combination for an array function).

*STEP* Select cell range L1:M20, click in the formula bar and enter the formula, *=SAMPLE(A1:A1000,J1:J1000,20,1)*, and press ctrl-shift-enter.

The numbers 0.001 appear in column L and numbers from 1 to 1,000 appear in column M. Every time you press F9, the values in column M change because you are drawing a new set of 20 from the population in column A. Column L stays the same because the probabilities of selection are the same for everyone.

*STEP* Create a histogram of the sample data. Make the bin widths 1 and press F9 until you see a taller bar than the others.

The taller bar signals an individual that appears twice in the sample. This is because we are sampling with replacement. Most surveys sample without replacement meaning an individual can be selected only once.

*STEP* Compute the sample average in cell N1 and enter the label, "sample average" in cell O1.

This is our estimate of the population average. Taking 20 draws, with replacement (putting each one back after we take it out), and computing the average of the 20 numbers is our DGP.

An estimate is a single number (point estimate) or range (interval estimate) produced by an estimator—a specific kind of statistic that is used to make a guess about a parameter.

Our sample average estimator is a random variable. It produces estimates (actual numbers) that bounce every time you hit F9. Is the sample average estimator any good? Let's find out.

*STEP* Click in the *Add-ins* tab and click *MCSim.* Run a 10,000 repetition simulation of the sample average (cell N1).

Your results will be similar to Figure 11.1, which is the sampling distribution of the sample average. The average of 10,000 sample averages of 20 observations will be be quite close to 500.5, the population average. The average displayed by the MCSim add-in is an approximation to the exact EV, which is 500.5.

Because the EV equals the population average, we say that the sample average estimator is *unbiased.* This is, of course, a desirable property. It is equivalent to *accuracy.* It means that the sample average estimator is centered on the truth.

| Summary Statistics | | Notes |
|---|---|---|
| Average | 500.229 | Exact EV = 500.5 |
| SD | 64.9964 | Exact SE = 64.55 |
| Max | 742.250 | |
| Min | 230.800 | |

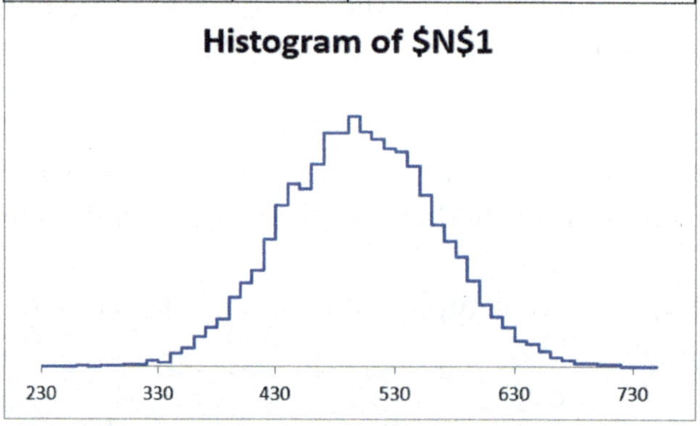


Figure 11.1: The sampling distribution of the sample average with n = 20.

But what about the *precision* of this estimator? Accuracy is good, but we also want precision. Think of throwing three darts—there are two ways you can do badly. First, you can be inaccurate, your three darts end up tightly clustered, but not on the bullseye. Another way to fail is that your throws are centered on the bullseye, but far apart. We want the darts clumped together *and* on the bullseye.

Accuracy is unbiasedness: on average, the estimator gives the truth. Precision is measured by the SE, the dispersion of the sample average statistic. The lower the better.

Figure 11.1, and your results, show that the SD of 10, 000 sample averages is about 65, which is close to the exact SE of 64.55. This is the typical bounce in the sample average.

Is this level of precision good? Well, we cannot prove it right now, but in the future we will see that there is no linear unbiased estimator with a lower SE than the sample average so this is the best we can do.

The exact SE of the sample average can be computed by a formula:

$$\text{Exact SE of the sample average} = \frac{\sigma}{\sqrt{n}}$$

Sigma is the population SD. We know it because we know the population values and computed it in cell B2.

STEP In cell Q1, compute the exact SE of the sample average, and then press F9 a few times.

The exact SE does not bounce. It is not a random variable.

The third SE, the estimated SE, is produced with sample information alone. In a real-world setting, we would not know the population (or why would we be sampling?) so we could not compute the exact SE of the sample average and we would not be repeatedly resampling many times to produce an approximate SE of the sample average. We would have one single sample.

With the data in our precious, single, simple random sample we could compute an SD. In fact, we could compute two different versions of the SD.

STEP In cell N2, enter the formula *=STDEV.P(M1:M20)* and the label "STDEV.P" in cell O2. In cell N3, enter the formula *=STDEV.S(M1:M20)* and the label "sample SD" in cell O2.

The two SDs are slightly different. The sample SD is an adjusted version of the population SD.

STEP In cell N4, enter the formula *=SQRT(20/19)*N2* and the label "sample SD by formula" in cell O3.

The two sample SDs are the same, which shows that the sample SD includes a correction factor, $\sqrt{\frac{n}{n-1}}$. What is being corrected for?

STEP Sim both the population and sample SDs (cells N2 and N3).

It is easy to see that the STDEV.S (the sample SD) is a better estimator than STDEV.P. We are trying to estimate $\sigma$, the SD of all 1,000 population values, the population SD in cell B2 and the sample SD's sampling distribution is centered more closely to $\sigma$ than the other SD.

We use our sample SD estimator to estimate the SE of the sample average via the same formula as before:

$$\text{Estimated SE of the sample average} = \frac{\hat{\sigma}}{\sqrt{n}}$$

The only difference in the formulas of the exact and estimated SEs is that one uses STDEV.P on the population data, $\sigma$, while the other uses STDEV.S on the sample data. But this has a big effect.

*STEP* In cell Q2, compute the estimated SE of the sample average, and then press F9 a few times.

The estimated SE, unlike the exact SE, is a random variable. It is built on sample data that changes every time we take a new sample. Is the estimated SE of the sample average a good estimator of its exact SE? We can run a sim to find out.

*STEP* Use the MCSim add-in to track cell Q2. Remember to remove the cell address in the second cell (the add-in remembers previous choices).

The results look pretty good. In fact, the estimated SE is just a little biased, but this goes away as the sample size increases. Thus, it is asymptotically unbiased and said to be a *consistent* estimator.

There are many paths to explore from this point in terms of statistical theory. We could examine the effects of sampling without replacement, varying the sample size, making the probabilities of selection unequal, and so on. We would be tweaking the DGP and this would affect the sampling distribution of any statistic produced by the sample data.

Instead of pursuing these theoretical issues, we return to our earlier CPS ASEC data to apply our new knowledge of the sample average as an estimator of the population average.

## 11.2 Estimated SEs in the Real World

We will assume the CPS is a simple random sample even though we know it is not because we only know how to estimate the SE for the sample average from an SRS. In fact, our SEs will be a little too small.

Our research question is this: What is the average income of workers ages 25 to 64 years old and how much does it vary from state to state? You might expect incomes to be roughly the same since people can move easily across states, but is this true?

If we had a census, a complete list of each worker's income, we could compute the exact answer for each state and the United States as a whole. But we do not. Instead, we have a random sample from the population.

*STEP* Return to IPUMS CPS and revise your extract. Include STATEFIP and download the data. Open the data in Excel and remove those with INCWAGE = 0. Use VLOOKUP to create a new variable called State that is the actual name of the state. Create a PivotTable that shows, for each state (including the District of Columbia), the number of observations (count), average INCWAGE, and the SD of INCWAGE.

Notice that you have more observations than you did before. This is because you only removed people with zero INCWAGE.

We know we are dealing with a sample drawn from a population. The average INCWAGEs (and SDs) in the PivotTable are random variables. If we sampled again, we would get different people and different sample statistics. What is the bounce in the average INCWAGE for each state?

*STEP* Add a column to the PivotTable that computes the estimated SE for average INCWAGE.

The results for all 50 states and DC are too large to display, but Figure 11.2 shows the first few rows of the table for 2019 ASEC data. Your results will be different because you are not using the 2019 CPS, but the pattern should be the same —California should have the highest count and lowest estimated SE.

| State | Count of INCWAGE | Average of INCWAGE | StdDev of INCWAGE | Estimated SE |
|---|---|---|---|---|
| Alabama | 1242 | $ 53,300 | $ 64,275 | $ 1,824 |
| Alaska | 795 | $ 60,067 | $ 68,227 | $ 2,420 |
| Arizona | 1120 | $ 56,845 | $ 58,539 | $ 1,749 |
| Arkansas | 1165 | $ 50,878 | $ 53,628 | $ 1,571 |
| California | 6943 | $ 63,786 | $ 75,553 | $ 907 |
| Colorado | 908 | $ 66,091 | $ 67,467 | $ 2,239 |
| Connecticut | 719 | $ 71,023 | $ 105,677 | $ 3,941 |

Figure 11.2: Average INCWAGE and its Estimated SE for a few states.

Consider Alabama in Figure 11.2. What does the estimated SE of $1,824 mean? This is our estimate of the bounce in the sample average INCWAGE. Even though we do not have an F9 key to press, we know that if we redid the survey, we get different people with different incomes and, thus, a different average INCWAGE. The estimated SE tells us how much variability we would get in the sample average: for Alabama, about $1,800.

The sample average is a point estimate of the unknown parameter, the population average, and the estimated SE of the sample average indicates the bounce so we would say, "Alabama's 25 to 64 year old workers had an average income of $53,300 give or take $1,800."

We will work on this in the future, but it is easy to see that we cannot conclude that Alaska has a higher average income than Alabama or Arizona because the bounce in the sample averages are so big.

On the other hand, even though we understand that the sample averages are random variables with baked-in bounce, some of the gaps between states are really big—too big to be explained by sampling variability. There really are differences in state-level average income across the United States.

The fundamental point is that the Average of INCWAGE values in the PivotTable are random variables. We are looking at one outcome of the chance process that produced the data.

## 11.3 Takeaways

An *estimator* is a recipe which tells us what do with the data; an *estimate* is a single outcome produced by the estimator. A single estimate is like rolling two dice and getting a one and a six. That is just one realization —roll again and you almost certainly get different numbers.

Accuracy and precision are not the same thing. Accuracy is about being centered on the truth, being correct on average, and is connected to unbiasedness. Precision is about dispersion and the clustering of the estimates—the lower the SE, the better.

We want an unbiased estimator that is also really precise. This way, we are likely to get an estimate near the truth.

An SE is a special type of SD. An SE tells you the dispersion of a statistic's sampling distribution. It is confusing, but it helps to remember that the term SE applies only to a statistic in an inferential setting—an SE tells you the bounce in a statistic that changes every time you draw another sample.

There are three SEs: exact, approximate, and estimated. In the real-world, we have access to only the last one.

To use statistical language correctly, always remember to fill in the blanks in "_____ SE of _____." The first blank is exact, approximate, or estimated. The second is the name of the statistic.

For example, there is an Exact SE of the sample average and an Approximate SE of the max streak and an Estimated SE of California's average income.

The SE formula depends on the statistic being analyzed. Sometimes, the formula is unknown (technically, we say there is no closed-form solution), but its SE exists and simulation can give us an approximate answer.

The estimated SE of the sample average, like the sample average itself, is a random variable and has a sampling distribution.

In papers, presentations, and conversation, we often drop the word estimated and speak only of the SE. If an adjective is not used, you can be almost certain that it is an estimated SE.

In earlier work, we brushed aside the distinction between population and sample SDs. Now that we understand how a sample is drawn from a population, we can explain the difference between the two. The sample SD is an adjusted version of the population SD that makes it a better estimator of the SD of the population values.

When dealing with a sample estimate of a population parameter, we often report the result as point estimate $\pm$ estimated SE.

Incomes in the United States exhibit wide spreads both within and across states. State-level differences in income are a bit of a puzzle since people can move. The CPS asks about where the respondent lived a year ago so it can be used measure migration.

A data generation process is an analogy between a chance process in the real world and a mathematical model. Free throw shooting is like randomly sampling from a box with two tickets (with 0 and 1). Rolling a die is like randomly sampling from a box with six tickets (numbered one to six).

For the statistical concepts used here (and in the rest of this book) to apply, the data generation process must be based on random sampling. This cannot be merely assumed, you have to know the process that generated the data and it must meet specific requirements. If not, it is, as they say, garbage in and garbage out.

# 12. Unemployment and its SE

The primary function of the CPS is to estimate unemployment rates for the economy as a whole, specific areas (such as major cities), and individual demographic and educational groups. These are reported every month as point estimates, but did you know that they have SEs? After all, they are sample-based statistics so they must be random variables.

Our goal is to see how the unemployment rate is computed and that it changes every time we take a new sample. We will implement a hypothetical data generation process in Excel and use it to run simulations.

Before we begin, we use the FRED Excel add-in to get the unemployment rate. FRED, the Federal Reserve Economic Data archive, compiles information from many sources. It is freely available online at fred.stlouisfed.org/, but we will access it through the FRED Excel add-in. This saves a lot of time and effort in getting the data in Excel for further analysis.

We bypass the appstore FRED add-in from fred.stlouisfed.org/fred-addin/ because it is error-prone and unstable with older Excel versions. Instead, we will work with a legacy FRED add-in file that is reliable and easy to use.

*STEP* Download *FRED.xlam* from dub.sh/legacyfred. Use the Add-ins Manager (keyboard shortcut *alt*, *t*, *i*) to install it.

*STEP* Enter "unrate" in cell A1, click *Get FRED Data* in the *FRED* tab. and make a FRED-style chart with shaded recession periods by clicking *Build Graph*.

These unemployment rates are point estimates in percentage terms. They move with the business cycle over time and the covid-19 pandemic led to a sharp increase as society locked down. In a well-functioning economy, unemployment rates in the range of 4 to 5% are considered healthy.

*STEP* Click on the link in column A to visit FRED's webpage for this variable.

The documentation shows that FRED (a data aggregator) got unemployment data from the Current Population Survey. This is the source of many labor market indicators in the United States.

FRED's webpage says nothing about the SE. This is common in media reports each month. The number is announced with no ± attached to it.

After we implement the DGP in Excel and understand how unrate is computed, we will reveal the Estimated SE of the unemployment rate, but you can take a stab at it right now.

*STEP* Fill in the blanks in this sentence: My best, most thoughtful guess of the Estimated SE of the unemployment rate is _._% points.

## 12.1 The Population

The CPS asks a series of questions to create mutually exclusive categories. It all starts with age. We will use the ROUND function with RAND to produce integer values of age from 1 to 100.

*STEP* Insert a sheet in the workbook. Create a population of 400 people ages 1 to 100 using the formula *=ROUND(RAND()*100 + 0.5, 0)* in cells A2:A401. Add the label "Age" in cell A1.

We can check our age distribution formula by creating a histogram.

*STEP* Make a histogram of age and change the bin width to 1. Press F9 a few times.

The gaps with no bars are when RAND() produces no one that age. We need to deaden these live values to have a fixed, unchanging population from which we will randomly sample.

*STEP* Copy range A2:A401 and then paste values.

Next, we separate the children, ages 15 and younger, from the adults.

*STEP* Make a dummy variable called Adult in column B for those 16 and older.

Now we need to use RAND() again to have some of the adults be in the labor force and others not. Roughly 60% of adults are in the labor force so we will use 0.6.

STEP In cell C1, enter the label "In LF" and in cell C2 enter the formula *=IF(B2=1, IF(RAND()<0.6,1,0), "")*. Fill down to cell C401 and deaden by copying and pasting values.

Finally, we need yet another dummy variable to make some of the people in the labor force be without work, yet searching for a job—they are unemployed.

STEP In cell D1, enter the label "Unemployed" and in cell D2 enter the formula *=IF(C2=1, IF(RAND()<0.1,1,0), "")*. Fill down to cell D401 and deaden by copying and pasting values.

These columns implement the basic layout of the CPS labor market survey. The questions are designed to separate adults from children, then to group adults in the labor force versus those not working or looking for work, and finally to group those adults in the labor force who are working from those not. This last category are unemployed.

Now we can compute our labor force statistics for the population. These are our parameters.

STEP Compute how many adults you have in cell G2, how many people are in the labor force in cell H2, and how many are unemployed in cell I2. Add labels for these in the first row.

The two primary labor market statistics are the *labor force participation rate*, the percentage of adults in the labor force, and the *unemployment rate*, the percentage of people in the labor force that are unemployed.

STEP In cell H3 enter the label "LFPR" and in cell H4 enter the formula *=H2/G2*. In cell I3 enter the label "unrate" and in cell I4 enter the formula *=I2/H2*. Format both cells as % with one decimal place.

Your population parameters, LFPR and unrate, should be around 60% and 10%, respectively. These are the stationary targets (they do not bounce) which we will use to evaluate our sample estimates.

## 12.2 The Sample

We will take a simple random sample of observations (n=50) from the population (N=400), where each individual is equally likely to be chosen, and explore sampling with and without replacement.

*STEP* In cell E1 enter the label "Prob selection" and in cell E2 enter 0.0025. Fill it down.

Now we are ready to use the SAMPLE function in the MCSim add-in. As you recall, it takes four arguments and the proper range of cells must be selected before you enter the formula. Since it is an array formula, it requires the ctrl-shift-enter keyboard combination to work properly.

*STEP* Enter the label "Prob selection" in cell L1. Copy cells A1 to D1 and paste in cell M1. Now we use the SAMPLE function. Select the range L2:P51. Enter the formula *=SAMPLE(A2:D401,E2:E401,50,1)* and press ctrl-shift-enter.

Next, we need to compute our sample estimates of their population analogues. We are most interested in the unemployment rate, but we could estimate any of the five variables in cells G1:I4.

*STEP* Copy cell range G1:I4 and paste in cell S1. Press F9 a few times.

The sample unemployment rate is in cell U4 and it changes every time you press F9. This is the telltale signal that it is a random variable. What is the expected value (center), SE (spread), and shape of the sampling distribution of the sample unemployment rate?

*STEP* Run a Monte Carlo simulation with the MCSim add-in that tracks cell U4. Do 10,000 repetitions.

Your results will be similar to that of Figure 12.1. The average of the sample unemployment rates will be close to the population parameter which is evidence that the sample unemployment rate is an unbiased estimator.

The approximate SE of the unemployment rate (the SD in the sim results) is about 6% points, which is quite large since this means we can easily get estimated unrates from 4% to 16%. Our estimates are unbiased, but so imprecise as to be almost worthless.

| Summary Statistics | | Notes |
|---|---|---|
| Average | 0.100 | |
| SD | 0.0615 | |
| Max | 0.368 | |
| Min | 0.000 | |

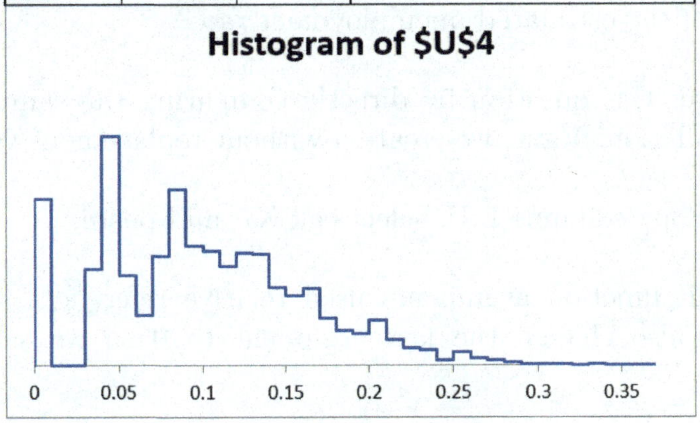


Figure 12.1: The sampling distribution of the estimated unemployment rate.

Notice also that the shape of the distribution is not close to the normal curve. There is a tall bar at zero (we often get no unemployed people in a sample because they are not that common in the population) and it is has a long tail.

Perhaps you are thinking that we should compute the exact SE of the unrate. That is something beyond the scope of this book because the unemployment rate is a complicated recipe.

We certainly cannot merrily compute $\frac{\sigma}{\sqrt{n}}$ and announce we have the exact SE. That formula works only for the sample average. The unemployment rate is not an average. It is a ratio—the number of unemployed (a sum) divided by the number in the labor force (a sum). Both the numerator and denominator are random variables.This is nowhere near the recipe for the average—add up the numbers and divide by n.

There is an analytical way to compute the exact SE of the unrate and corresponding estimated SE, but we do not need it. We have simulation to give us a good idea of the bounce in the sample unrate. The SD of our sim is an approximate SE.

Is there anything we can do to improve the precision of our unrate estimator? The next section explores two possibilities.

## 12.3 The Effects of Replacement and Sample Size

In the real world, you cannot be surveyed more than once. The CPS samples without replacement. What effect does this have on the sampling distribution of the estimated unemployment rate?

We can answer this question by directly comparing the sampling distributions via simulation. First, we create a without replacement version.

*STEP* Copy columns L:U, select cell X1, and paste.

The SAMPLE function arguments used relative references so we have to fix that and also change the last argument to 0 so we sample without replacement.

*STEP* Select cell X1 and click in the formula bar. Make the formula *=SAMPLE(A2:D401,E2:E401,50,0)* and press *ctrl-shift-enter*.

We can now run simulations of both unrates, with and without replacement, to see the difference.

*STEP* Use the MCSim add-in to track both cell U4 and AG4 with 10,000 repetitions.

You will find that both are unbiased, but estimated unrate without replacement has a slightly lower SE of the unrate than with replacement. This is due to the fact that the population is shrinking as individuals are selected.

Statisticians have worked out a correction factor (CF) for the SE without replacement:

$$CF = \sqrt{\frac{N-n}{N}}$$

*STEP* In cell M5 of *MCSim* sheet, compute *=SQRT((400-50)/400)*. In cell M6, enter the formula *=M5*J6*.

As advertised, the SE of the unrate without replacement is about 93% of its with replacement counterpart.

This is a good result since it means that we are getting a more precise estimator. However, before you get too excited about this, consider that the CPS is sampling 60,000 households out of a population of 130 million households. Thus, the correction factor is 0.9998. Since the sample size if teeny relative to the population, the correction factor is almost one. The fact that we are sampling without replacement is not going to help us lower the SE of the unemployment rate by much.

What about changing the sample size? We can double n from 50 to 100 to see what this does to the SE.

*STEP* Copy columns L:U, select cell AJ1, and paste.

Once again, we have to correct the SAMPLE function arguments. We will use with replacement since we know that it is of little consequence in the real world.

*STEP* Select cells AJ2 to AN101 (since n = 100) and click in the formula bar. Make the formula *=SAMPLE(A2:D401,E2:E401,100,1)* and press ctrl-shift-enter.

Since we are going to race these two estimators against each other, we can make the display easier to read by naming cells before we sim.

*STEP* Click on cell U4 and name it "unrate50" and then click on cell AS4 and name it "unrate100". Sim cells U4 and AS4 with 10,000 repetitions.

Figure 12.2 shows the promising results. The approximate SE of the unemployment rate, which is a reliable measure of the exact SE, is about 2% points and a third lower with n=100 than with n=50.

| unrate50 | | unrate100 | | Notes |
|---|---|---|---|---|
| Average | 0.100 | Average | 0.100 | |
| SD | 0.0630 | SD | 0.0434 | |
| Max | 0.471 | Max | 0.310 | |
| Min | 0.000 | Min | 0.000 | |

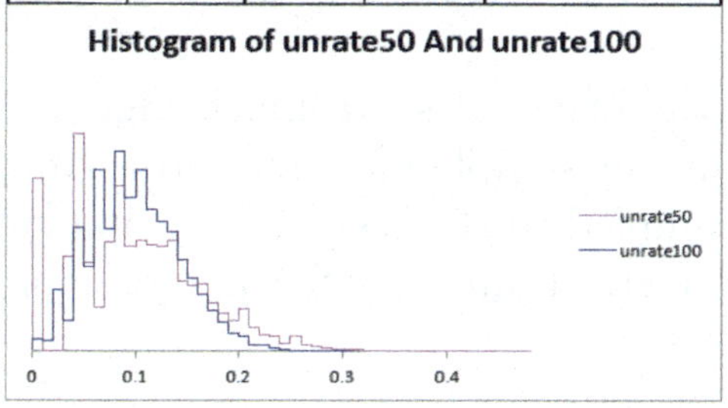


Figure 12.2: Increasing n lowers the SE.

Increasing the sample size definitely lowers the SE. although doubling n did not cut the SE by half. Of course, it is still way too big to be useful, a ± of over 4% points is still way too much bounce in the estimate.

But remember that the CPS is polling 60,000 households with multiple people in each household. How big is the SE?

## 12.4 The SE of Unemployment Rates

The Bureau of Labor Statistics publishes detailed technical documentation. Figure 12.3 reproduces Table 1-C, which shows SEs for various groups, from *Technical Notes* (www.bls.gov/cps/eetech_methods.pdf.)

Table 1-C. **Approximate standard errors for unemployment rates by major characteristics**

(In percent)

| Characteristic | Monthly rate | Consecutive month-to-month change |
|---|---|---|
| Total | 0.10 | 0.12 |
| Men | .14 | .16 |
| Men, 20 years and over | .14 | .15 |
| Women | .14 | .16 |
| Women, 20 years and over | .14 | .16 |
| Both sexes, 16 to 19 years | .78 | .97 |
| White | .11 | .12 |
| Black or African American | .44 | .49 |
| Asian | .46 | .51 |
| Hispanic or Latino ethnicity | .34 | .38 |
| Married men, spouse present | .14 | .15 |
| Married women, spouse present | .16 | .18 |
| Women who maintain families | .46 | .52 |

Figure 12.3: SEs for various unemployment rates.

The circled number in Figure 12.3 is the answer to our question—this is the bounce in the national unemployment rate that is reported every month. It is small, 0.1% point, because the sample size is pretty large.

Notice how some of the other SEs are much bigger. This is because they are based on much smaller sample sizes for those groups. In particular, not many teenagers are sampled so the teenage unemployment rate (Both sexes, 16 to 19 years) has an SE of almost 0.8% points—that is a lot of sampling variability.

So what does the SE of the national unemployment rate being 0.1% tell us? Well, for one thing, when you hear that the unemployment rate fell last month from say 5.6% to 5.5%, this is meaningless. We cannot conclude that the unemployment rate in the population fell because we do not have the resolution to see that clearly. It will take moves of 0.2 percentage points or more before we can say things are really changing.

If you hear an unemployment rate for a city or a very specific group, you can be sure that its SE is much higher. You should be careful in interpreting a change and remember that sampling variability is baked into sample statistics.

## 12.5 Takeaways

The unemployment rate is estimated every month with data from the CPS.

The survey is designed to create a series of mutually exclusive categories that give us the number of adults in the labor force and then we divide those into employed and unemployed (actively looking for work).

The unemployment rate is the ratio of unemployed to the total labor force (which is employed plus unemployed).

The LFPR is another ratio-type statistic that is commonly used. It is the number of people in the labor force divided by the number of adults.

Since unrate is computed from a sample, it has an SE and it is roughly 0.1% for the overall, national unemployment rate.

The unrate's SE is reasonably precise because the sample size is relatively large.

Sampling without replacement does not matter much for the unrate's SE because the sample size is really small compared to the population. The correction factor can be used to compute the effect of sampling without replacement.

There are formulas for the unrate's SE, but we can use simulation to give us a close approximation of the exact SE.

[illegible]

[illegible]

[illegible]

[illegible]

[illegible]

[illegible]

[illegible]

[illegible]

[illegible]

[illegible]

[illegible]

[illegible]

# 13. Search Theory

Economists usually assume perfect information: buyers choose how much to buy based on preferences, income, and given prices. We know, however, that buyers do not face a single price for a product—there is a distribution of prices and sellers change their prices frequently.

In the paper that started the subfield of search (or information) economics and won him a Nobel Prize, George Stigler wrote, "Price dispersion is a manifestation—and, indeed, it is the measure—of ignorance in the market" (Stigler, 1961, p.214).

You would think consumers would be unable to choose in such an environment. After all, how can they know the budget constraint without prices? The answer is that they search or, in other words, they go shopping, and then use the lowest prices found to solve their constrained utility maximization problem.

*Search Theory* is an application of the economic approach to the problem of how long to shop in a world of many prices. Search is a productive activity because it enables one to find lower prices, but it is costly. One can search too little, and end up paying a higher price; or one can search too much, and spend hours only saving a few pennies in the end.

Our goal is to set up and solve a search optimization problem with simulation. We will continue our focus on unemployment by modeling a worker searching for a job. The idea is that she decides in advance how many interviews and wage quotes to obtain, then picks the best offer. We seek the optimal number of jobs to sample.

After we find the solution, we explore how search costs affect the time spent searching, during which one is unemployed. The internet, for example, should lower search costs and this should affect how long workers search and, therefore, the unemployment rate.

Along the way, we will see that the MCSim add-in comes with more than just the SAMPLE function. We add UNIFORM and NORMALRANDOM to our stock of Excel functions.

## 13.1 Describing the Search Optimization Problem

We assume that workers do not know the wages for each job before interviewing and getting a job offer. We simplify the problem by assuming that the jobs in different firms are identical (i.e., homogeneous) so the worker just wants to get highest wage. Unfortunately, applying and interviewing is costly so the worker has to decide how many offers to obtain.

The first task is to create the distribution of wages faced by the worker. We have been using Excel's native RAND() function, but the MCSim add-in comes with its own uniform distribution function. UNIFORM has two arguments, mean and SD. We can compare the two to confirm they are the same.

*STEP* In cell A1 of a new workbook, enter the formula *=RAND()*. In cell A2, enter the formula *=UNIFORM(0.5,1/SQRT(12))*. Press F9 a few times.

We use $\frac{1}{\sqrt{12}}$ because this is the SD of a uniformly distributed random variable on the interval from 0 to 1 (like RAND()). Pressing F9 shows both bounce, but does not tell us much more than that. We can run a simulation to approximate their distributions.

*STEP* Use the MCSim add-in to track both cells with 10,000 repetitions.

Notice that the numbers are not exactly evenly distributed. They are drawn from a uniform distribution, but each realization (like the roll of a die) deviates from a purely rectangular distribution due to randomness in sampling from the uniform distribution. The more repetitions you run, the closer the simulation's histogram will get to a smooth, rectangular distribution.

The results demonstrate the two functions are similar. The advantage of the UNIFORM function is that we can easily set the mean and SD. A disadvantage is that it not native so it will not work without the MCSim add-in and it is slower than RAND().

*STEP* Insert a sheet in the workbook and name it "DGP."

We will use the first five rows for parameters. We will begin with a population of 100 jobs and their wages. Then we will sample from the population as we model the search process.

*STEP* In cell A6, enter the label "Job" and create a series from 1 to 100 starting in cell A7. In cells A2 and A3, enter the labels "mean" and "SD." In cell B1, enter the label "Uniform." In cell B2, enter the value 0.5 and in cell B3, enter the formula *=1/SQRT(12)*. In cell B6, enter the label "Wage" and in cell B7 enter the formula *=UNIFORM($B$2,$B$3)*. Fill it down. In cell A5 enter the label "pop max" and in cell B5, enter the formula *=MAX(B7:B106)*. Format column B as $.

This is our population of jobs and their wages. Workers know the distribution of wages, but they do not know which job has which wage until they interview, so they cannot immediately go to the firm that has the highest wage and take that job. Instead, this search model says that the worker chooses a number of jobs to interview for (that is, sample) and then chooses the highest of the observed wages.

## 13.2 Searching as Sampling

There is one more piece of information we need before we can sample: the probabilities of selection. We assume a simple random sample so all jobs are equally likely to be chosen.

*STEP* In cell D6 (we skip column C), enter the label "prob selection" and the value 0.01 in cell D7. Fill it down.

We begin by choosing three jobs to search via the SAMPLE function. We are not sure this is the optimal number of jobs to interview and get wage offers, but we can see how it does and then compare it to the outcomes of other number of jobs searched.

Recall that the SAMPLE function has four arguments, the population range, the probabilities of selection range, the sample size, and an optional with or without replacement choice. Since we will refer to the analytical solution shortly and it assumes sampling with replacement, this is what we use.

*STEP* In cell G6, enter the label "prob selection", in cell H6, enter the label "Job", and in cell I6, "Wage." Select cell range G7:I9, click in the formula bar, enter the formula *=SAMPLE($A$7:$B$106,$D$7:$D$106,3,1)*, and press ctrl-shift-enter. Copy cell B5 and paste in I5, enter the label "sample max" in cell H5, and then format column I as $. Press F9 a few times.

The concept of job search is quite clear: it is like sampling from a population of offers and then we take the biggest one (the max in cell I5). Intuitively, three jobs seems too few because we are unlikely to get a really high paying offer. But before we proceed further, what about the costs of searching?

*STEP* In cell H1, Enter the label "cost per offer" and in cell I1 enter the value 0.02.

Let us say each job offer costs 2 cents. We need to multiply cost per offer by the number of jobs we decide to sample. Since we will be changing this, it is convenient to have Excel compute this for us.

*STEP* In cell H2, enter the label "#offers" and in cell I2 enter the formula *=COUNT(I7:I106)*. Change the formatting to General (it is 3 jobs so dollar units are incorrect).

The COUNT function reports the number of non-blank cells so it displays 3 in cell I2.

*STEP* In cell H3, enter the label "search cost" and in cell I3, enter the formula *=I1*I2*.

Cell I3 reports the worker's total search cost and it represents the negative side of search. More search yields a higher max wage offer, but this is not free. Three job applications and interviews may not be good at getting a high wage offer, but the good news is the worker only spends 6 cents on that sample.

*STEP* In cell H4, enter the label "net wage" and in cell I4 enter the formula *=I5-I3*.

Cell I4 is the value we are interested in maximizing. Net wage is a random variable because the highest offer in the sample is a random variable. What is the expected value, spread, and sampling distribution of net wage? We simulate to answer this question.

*STEP* Use the MCSim add-in with 10,000 repetitions to evaluate searching with three job offers. Upon completion, enter "n = 3" in the Notes area of the simulation output to describe this run.

Figure 13.1 shows results; yours will be a little different, but we can say that the approximate expected value is about \$0.69. This is the typical net wage the worker will receive. The approximate SE of the sample net wage is about 20 cents and the sampling distribution is obviously not normal.

As in previous work on estimating unemployment rate, the exact SE of the sample net wage is not simply $\frac{\sigma}{\sqrt{n}}$ because the net wage statistic is not the sample average. Deriving the formula for the exact SE of sample net wage is beyond us and not needed, since simulation gives us a reliable measure of it.

| Summary Statistics | | Notes |
|---|---|---|
| Average | 0.687 | n = 3 |
| SD | 0.1939 | |
| Max | 0.940 | |
| Min | -0.033 | |

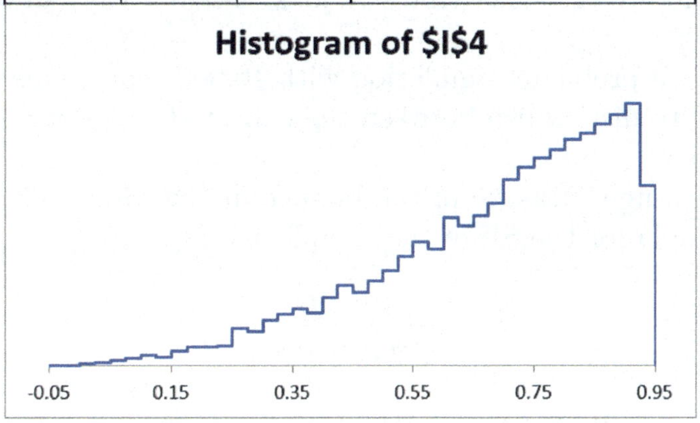


Figure 13.1: The sampling distribution of net wage with three jobs.

Figure 13.1 shows there is a great deal of variability in the net wage. We often get a high wage in the sample so we end up with a net wage of \$0.75 or higher, which is pretty good. Less frequently, but it does happen, none of our three job offers are good so we end up in the bottom left corner of the sampling distribution. In fact, the minimum and a few other values are negative which means the max wage in the sample of three was less than the 6 cents of search cost!

*STEP* Scroll right to see the values on which the chart is based. You can see exactly how many there are in each bin.

Perhaps the worker would do better by sampling more jobs? Yes, the search cost would increase, but maybe the higher wage we would expect to get (since we are sampling 6 jobs instead of 3) would more than compensate for the higher search cost?

*STEP* Select cells G7:I12, change the 3 to a 6, and press ctrl-shift-enter.

The COUNT function automatically updates our search cost cell and we are ready to simulate.

*STEP* Run 10,000 repetitions with cell I4. What do you find?

Your simulation results should show an approximate expected value of about $0.73, which is better than our first result. It is true that the gains from more search outweigh the costs, but is 6 the optimal solution?

*STEP* Modify the sheet to test 7 jobs. Is 7 better than 6?

Sadly, we have a problem: simulation with 10,000 repetitions does not have the resolution to distinguish between the expected value of 6 versus 7 jobs.

We can get a rough measure of the bounce in the Monte Carlo average by using the formula for the SE of the sample average:

$$\frac{simSD}{\sqrt{repetitions}}$$

*STEP* In a cell H5 of your *MCSim* sheet, compute the bounce in the simulations's average by entering the formula, *=J6/SQRT(F3)*.

The simulation's average of 10,000 net wages can be expected to vary by about 0.001. That seems quite precise, but it is not precise enough to show the difference in the expected value of net wage for 6 versus 7 jobs.

We could try to run more repetitions, but some quick analytical work will show that it is going to take a huge number of repetitions for simulation to give us a reliable answer.

This way of measuring the bounce in the sim average is confusing. It is *not* the dispersion of the sample statistic—that is given by the SD of the sim (the approximate SE of the sample statistic). Instead, it is quick way to measure how much the sim average will change if we sim repeatedly.

## 13.3 Analytical Methods

The optimal search optimization problem can be solved via analytical methods. For the uniform wage distribution on the interval from zero to one, the average maximum wage after obtaining $n$ job offers is:

$$Average\ Max\ Wage = \frac{n}{n+1}$$

*STEP* Run a simulation on any of the MAX cells in row 5 to verify this. Compute the formula (for the n jobs) and compare it to the average max wage from the simulation. They will be quite close.

The equation for the average max wage shows that it is increasing as $n$ rises and it does so at a decreasing rate. In other words, there are diminishing returns to searching for high wages.

The worker's optimization problem is to maximize the expected net wage, which is the Average Max Wage minus search costs, $cn$:

$$\max_n NetWage = \frac{n}{n+1} - cn$$

To find $n^*$, take the derivative with respect to $n$ and set it equal to zero:

$$\frac{dNetWage}{dn} = \frac{1}{n+1} - \frac{n}{(n+1)^2} - c = 0$$

$$\frac{1}{n+1} - \frac{n}{(n+1)^2} = c$$

This equimarginal condition says that the optimal sample size is found where marginal returns from additional search equals marginal cost (c). As long as the additional returns from obtaining one more wage offer exceeds the additional cost of one more interview, the worker will continue to search.

From the equimarginal condition, we can solve for optimal $n$ to get a reduced form solution.

$$\frac{1}{n+1} - \frac{n}{(n+1)^2} = c \rightarrow \frac{1}{(n+1)^2} = c \rightarrow \sqrt{\frac{1}{c}} = n+1 \rightarrow \sqrt{\frac{1}{c}} - 1 = n^*$$

With $c$ = \$0.02 per job, optimal $n$ = 6. Figure 13.2 makes clear that the objective function, net wage, is quite flat around its maximum. For simulation to distinguish between an expected net wage at n=6 of \$0.737 versus n=5 or n=7—a few thousandths of a cent less—would take much more than 10,000 repetitions.

| n | $\frac{n}{n+1}$ avg max wage | $0.02n$ search cost | net wage |
|---|---|---|---|
| 1 | \$ 0.50 | \$ 0.02 | \$ 0.480 |
| 2 | \$ 0.67 | \$ 0.04 | \$ 0.627 |
| 3 | \$ 0.75 | \$ 0.06 | \$ 0.690 |
| 4 | \$ 0.80 | \$ 0.08 | \$ 0.720 |
| 5 | \$ 0.83 | \$ 0.10 | \$ 0.733 |
| 6 | \$ 0.86 | \$ 0.12 | \$ 0.737 |
| 7 | \$ 0.88 | \$ 0.14 | \$ 0.735 |
| 8 | \$ 0.89 | \$ 0.16 | \$ 0.729 |
| 9 | \$ 0.90 | \$ 0.18 | \$ 0.720 |
| 10 | \$ 0.91 | \$ 0.20 | \$ 0.709 |

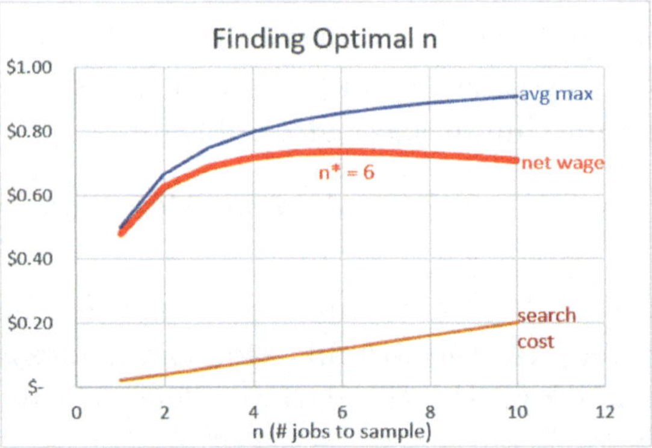


Figure 13.2: The Optimal Solution.

## 13.4 Comparative Statics

The availability of information is a key component of search cost. Informational advertising is how firms let workers know where they are and what jobs are open. We can model this type of advertising as a decrease in search costs—today, unlike in the past, workers can go online to gather information on which jobs are being offered and their wages. Search costs are still positive (workers do not know, for example, whether all firms advertise online or just some), but lower than without advertising.

The reduced form expression makes comparative statics analysis straightforward. If the internet makes it easier to search for jobs, so c falls to \$0.01 per job, then we can quickly compute that the optimal number of jobs is 9.

Figure 13.3 shows optimal n for per unit search costs from \$0.01 to \$0.10. As the internet emerged in the early 21$^{st}$ century, unemployment rates rose and stayed stubbornly high. Some economists believed search theory could explain what was going on. If per unit search costs fall, the optimal number of jobs sampled rises and this leads to higher baseline unemployment rates.

| c | Optimal n |
|---|---|
| $ 0.01 | 9 |
| $ 0.02 | 6.071068 |
| $ 0.03 | 4.773503 |
| $ 0.04 | 4 |
| $ 0.05 | 3.472136 |
| $ 0.06 | 3.082483 |
| $ 0.07 | 2.779645 |
| $ 0.08 | 2.535534 |
| $ 0.09 | 2.333333 |
| $ 0.10 | 2.162278 |

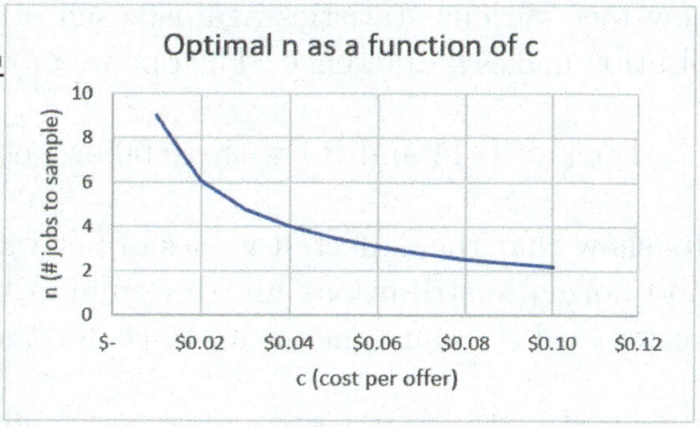


Figure 13.3: Comparative statics.

## 13.5 Uniform versus Normal

We close with a question that is quite difficult to answer with conventional methods, but relatively easy with simulation. It concerns the distribution of wage offers in the population. It is unlikely they are uniformly distributed. One possibility is that they are normally distributed, mostly bunched around the center with a few lower and higher than the mean.

The question is: How does the distribution of wages affect the sampling distribution?

*STEP* Copy columns A and B, select cell L1, and paste. Change cell M1 to “Normal”. Select cell M7 and replace the UNIFORM with the NORMALRANDOM function: *=NORMALRANDOM($M$2,$M$3)*. Fill it down.

The two distributions have the same means and SDs, but different shapes.

*STEP* Run a simulation of cells B7 and M7 with 10,000 repetitions to see the difference between the two distributions.

Next, we sample from the normally distributed population so we can compare it to sample statistics with a uniform distribution.

*STEP* Copy cells G1:I12, select cell P1, and paste. Edit cell P7 so it draws from the normally distributed population, using the formula: =SAMPLE($L$7:$M$106,$D$7:$D$106,6,1).

We can now race various statistics against each other to see if the population distribution makes a difference. The one we most care about is net wage.

*STEP* Track cells I4 and R4 with 10,000 repetitions.

The results show that the expected values of net wage are quite close so the uniform and normal distributions are the same in this respect. This makes sense since we used the same means and SDs for the two populations.

The SDs and the shape, however, are quite different. The normally distributed population produces a bell-shaped sampling distribution.

This is not to say that the normal distribution is correct for the population. Perhaps a lognormal (with a long tail) or some other distribution better describes wage offers. The point is that, in this case, the distribution of the population influences the sampling distribution.

## 13.6 Takeaways

With perfect information, consumers, workers, and firms know all prices and wages. Incomplete information complicates things. Because agents will not obtain every price, since that is too expensive, we see price dispersion. This is a major result of search theory and it deserves further explanation.

You would think that competition would tend to make prices of the same product equal. This is known as the *Law of One Price*. But this only applies to a world where agents can costlessly gather prices.

In other words, the Law of One Price will fail to hold whenever it is costly to collect price data. This is true in the real world, where some workers will end up getting higher wages than others because the maximum wage in their particular information set is different than the maximum wage in another worker's sample.

Because lower search costs induce more search, a reduction in search costs would have the effect of reducing (but not eliminating) wage dispersion. Because optimizing workers will choose not to canvass every firm for wage offers as long as search is costly, wage dispersion will exist. This is the key result of the fixed sample search model.

Economists have been interested in search theory for decades. The internet promised a big decrease in search cost and it may well have delivered that, but more recently, technology has really upended search theory. Today, your online search behavior is monitored and your clicks influence the prices you see.

The next level search models do not treat the population of prices as given and do not allow the consumer to randomly sample without changing the price distribution. Consumers still have an optimization problem to solve, but so do firms.

Taking the sim's SD and dividing by the square root of the number of repetitions is a handy way to get a sense of the bounce in the sim average, but it can be confusing. This approach is a rough measure of how much the sim average would change if we ran many more sims. It helps us figure out when a sim provides evidence of being close or far away from a specific value.

*EXCEL TIP* The SAMPLE array function (like BIVARNORMAL) is a powerful user-defined function in the MCSim add-in. SAMPLE has four arguments:

=SAMPLE(pop range, prob range, n, 0 or 1 replacement)

## References

Stigler, G. (1961) "The Economics of Information," The *Journal of Political Economy*, pp. 213–225, www.jstor.org/stable/1829263.

Stigler was trying to explain price dispersion, but search theory has expanded far beyond this and is especially important in Labor Economics. For a good review of search theory models, see Rogerson, R., Shimer, R., and Wright, R. (2005) "Search-Theoretic Models of the Labor Market: A Survey," *Journal of Economic Literature*, pp. 959–988, www.jstor.org/stable/4129380.

# 14. Midterm

It is time for a break!

*STEP* Visit dub.sh/regcourse to see what a midterm exam looks like for this course. Practice by taking the exam, guess how you did, then check your guess by grading yourself with the answer key. How did you do?

# 15. Measurement Error

The first data generation process—it was not called that by its creators—came from astronomy. The DGP was used to combine measurements from new devices called telescopes to estimate orbits. Scientists in the 17$^{\text{th}}$ and 18$^{\text{th}}$ centuries were baffled by a difficult puzzle: theory said that planets move along smooth curves, but their measurements did not produce perfect curves. They were squiggly, with tiny movements away from the predicted orbit.

An important advance was made when astronomers realized that the deviations were produced by an imperfect measurement process of the exact location of the planets. The noise was not in the planets, but produced by the telescope and measuring process itself.

The next challenge was to figure out how to combine the imperfect measurements. Astronomers knew that they should make use of all of the observations, but how? By explicitly laying out the data generation process, Gauss and others were able to prove that there is a best way to use the data.

Our goal is to show the DGP and make clear how it produces observed values which are the outcomes of a random variable. As you know, there are three SEs and this gives us an opportunity to review and cement core ideas.

We will see that we can use the same formula for the SE for the Polling and Measurement Error DGPs. We take the time and effort to differentiate between the two DGPs, even though they have the same statistical properties, to emphasize a fundamental lesson of data analysis: *we must have a coherent, plausible explanation for the data generation process before estimating parameters and applying the machinery of statistical theory.*

## 15.1 Implementing the Measurement Error DGP

We can describe the DGP of the Measurement Error model formally:

$$y_i = \mu + \epsilon_i, i = 1, ..., n$$

$$E[\epsilon_i] = 0, i = 1, ..., n$$

$$SD[\epsilon_i] = \sigma, i = 1, ..., n$$

$$\epsilon_i \text{ is distributed independently of } \epsilon_j$$

The first equation says that there is a true, exact value, $\mu$ (mu), but we cannot directly observe it. Instead, we see $y$ each time we measure which is composed of the true, exact value plus epsilon ($\epsilon$), which is a draw from an error distribution. Both $\mu$ and $\epsilon$ are invisible. We can only see $y$.

The next three equations describe the error distribution. First, the expected value of the errors, $E[\epsilon_i]$, is zero. This means the measuring process is unbiased. Sometimes the measuring noise adds to the true, exact value and other times it subtracts from it, but these positive and negative errors balance out so, on average, the error is zero.

Second, the errors are identically distributed with the same dispersion, $\sigma$ (sigma). This is *homoskedasticity.* It is important that $\sigma$ does not have a subscript. This means that the errors are being drawn from a distribution with the same spread.

The last equation says that the errors are independent of one another. This means that the value of one error has nothing to do with another error. So, say one measurement is off by a large, positive error, that has no effect on the next measurement.

The last two equations are often summarized by the phrase, *independently and identically distributed*, or even shorter, *iid*.

The philosophy of simulation is based on implementing the DGP and directly observing the results, so let's do that now. We begin with the parameter $\mu$.

As a concrete example, suppose that we are measuring the length of a board that is exactly 1234.56789 millimeters long with a ruler.

*STEP* In a blank workbook, name the sheet "DGP." In cell A1, enter the number 1234.56789 and widen the column so you can see the entire number. In cell B1, enter the label "mu" and name cell A1 "mu." In cell A2, enter the number 1 and name the cell "sigma." Add the label "sigma" in cell B2.

The parameter *sigma* reflects the precision of the measuring device. We can expect a 1 mm variation each time we measure the board.

The word *error* is misleading. There is *no mistake* when we measure. We are careful, but with the ruler we are using, we get slightly different measurements. In statistics, error is simply random noise.

If we had a laser measuring device, perhaps $\sigma$ would be 0.001 mm, which is a micron. This would, naturally, greatly reduce measurement error, but it would not eliminate it completely.

*STEP* In cell A4, enter the label "Observation" and enter the numbers 1, 2, and 3 in cells A5, A6, and A7. In cell B4, enter the label "epsilon" and in cell B5, enter the formula: *=NORMALRANDOM(0,sigma).* This function is in the *MCSim* add-in so install it (*alt, t, i*) if needed. Fill it down to cell B7. Press F9 a few times.

The fact that the three cells have the same mean and SD fulfills the "identically distributed" requirement of the DGP. The fact that one cell is in no way connected to the others means they are independent. Thus, our implementation is iid.

Cells B5, B6, and B7 produce random draws from a normal error distribution. It does not have to be normally distributed, but measurement errors do tend to follow the normal curve. We can confirm that our implementation is working as advertised via simulation.

*STEP* Sim cell B5 with 10,000 repetitions.

Your results should show an average of the 10,000 random numbers close to 0 with an SD close to 1. This is evidence that the NORMALRANDOM function is working properly. This is not, of course, a sampling distribution since there is no statistic being simulated.

The next step is crucial. We will combine $\mu$ and the random error value to produce observed y.

*STEP* In cell C4, enter the label "y" and in cell C5 enter the formula: *=mu+B5.* Fill it down. Press F9 a few times.

The key idea is that when we measure the board three times, we see the values in cells C5, C6, and C7, but we cannot separate the two components, $\mu$ and $\epsilon$. To emphasize this, we can use color.

*STEP* Select cells A1:B2 and change the color text to red. Make the text color of cells B4:B7 red.

Red text means we cannot directly observe these values. We do not know the true, exact length of the board; the true, exact precision of the measuring device; or the actual random errors drawn each time we measure.

All we can see is the number for y, which we know is composed of a deterministic part ($\mu$) and a stochastic disturbance term ($\epsilon$). We cannot disentangle the two components.

## 15.2 Estimating $\mu$

We can estimate $mu$ with our three observations by taking their average.

*STEP* In cell E4, enter the label "sample average" and in cell E5, enter the formula: *=AVERAGE(C5:C7)*. Press F9 a few times.

We will want to simulate cell E5, of course, but first let's create a graph that gives us another way of seeing what is going on here.

*STEP* Make a scatterplot of observed y. Select cells A5:A7, hold down the ctrl key, select cells C5:C7, and then choose a Scatter chart (no lines). Select the chart, click in the formula bar, and enter the formula: *=SERIES("mu",{1,3},{1234.56789,1234.56789},1)*. Click one of the dots in the series you just created, click *Chart Type* (in the *Design* tab), and change the type for the series to straight lines with no markers. Make its color red. Finally, double-click the $y$ axis and make the min and max ranges 1230 and 1240, respectively.

Now, when you press F9 repeatedly, the dots bounce around, but the horizontal line does not move. This is exactly what the DGP says: there is a constant, fixed parameter value, but the observations have built-in randomness from the measuring process.

Notice also how Excel accepts numbers in curly brackets for the SERIES function. This offers another way to add data to a chart.

*STEP* Add another series to your chart for the sample average. Enter this formula: *=SERIES(,(DGP!$A$5,DGP!$A$14),(DGP!$E$5,DGP!$E$5),1)*.

Note how this formula uses parentheses as an alternative to the usual cell range, such as A5:A14 (like the formula for the data). Press F9 a few times.

The sample average is bouncing around, above and below, $\mu$. We want to know its sampling distribution, including center, spread, and shape.

*STEP* Run a simulation with 10,000 repetitions of cell E5.

The average of the 10,000 sample averages is close to $\mu$, 1234.56789. This is evidence that the sample average estimator is unbiased.

The SD is interpreted as an approximate SE. The exact SE of the sample average for the measurement error DGP is given by:

$$\frac{\sigma}{\sqrt{n}}$$

*STEP* In cell L6, enter the formula: *=1/SQRT(3)*.

It is easy to see that the approximate SE from the simulation is quite close to the exact SE.

The shape of the distribution looks normal, but we can make a clearer visual display to illustrate that the sampling distribution follows the normal curve. We will compute the number of sample averages for each bin and add this to the chart.

*STEP* Scroll right and, in cell AM1 enter the label "N CDF." In cell AM2, enter the formula: *=NORM.DIST(AK2,$J$5,$J$6,1)*. Fill it down.

You have the *cumulative distribution function*, CDF, for each bin value in column AK. The CDF tells us the probability that a random variable is less than or equal to a given value. So, for a normally distributed random variable, the chances that it is less than or equal to zero, in standard units, is 50%.

To compute the number expected in each bin, we can take the difference in the CDF from one bin to the next and multiply by 10,000. The bins are repeated in column AK to draw the horizontal part of the bar in the chart so we have to account for this in our formula.

*STEP* In cell AN1, enter the label "N prob" and in cell AN3 (skip AN2), enter the formula: *=(AM4-AM2)*10000.* Fill it down to the penultimate cell, AN100. You should see pairs of repeated values.

Some of the pairs in columns AL and AM are quite close, while others are farther apart. Remember, simulation is an approximation and it will not give us exactly what we expect to see in each bin.

*STEP* Add the data in column AN to the chart by copying the existing SERIES formula and changing AL to AN.

The result is shown in Figure 15.1. This is clear visual evidence that the sample average of the measurement error DGP with normally distributed errors follows the normal curve.

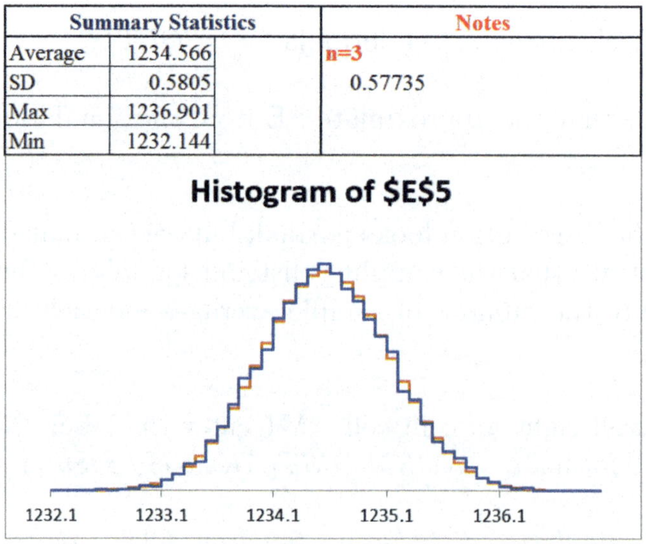


Figure 15.1: The sample average follows the normal curve.

## 15.3 Estimating the Exact SE

As you know, however, we would not have access to the exact SE or approximate SE of the sample average in a real-world situation. We do not know *sigma* so the exact SE cannot be computed. We also do not have the information needed to set up and run a simulation.

Thus, we are left with the third SE, the estimated SE, as the only practically obtainable measure of dispersion in the sample average. We use the same formula as the exact SE, except we estimate *sigma*.

The estimate of sigma, also known as sigma-hat, $\hat{\sigma}$, comes from the SD of the residuals. Recall that residuals are simply actual minus predicted values.

**STEP** In cell G4, enter the label "residuals" and in cell G5, the formula: *=C5-$E$5*. Fill it down to cell G7.

It is easy to see by looking at the graph and the values in cells G5:G7, that the residual is positive when actual (observed) y is above the sample average line and negative when below. Notice that the residual does not utilize $\mu$—the residual is computed from observed data.

**STEP** In cell H6 enter the label "SD of residuals" and in cell H7 enter the formula: *=STDEV.S(G5:G7)*.

We use the sample SD, STDEV.S, (the population SD multiplied by its degrees of freedom, $\sqrt{\frac{n}{n-1}} = \sqrt{\frac{3}{2}}$) to give us a better estimate of the SD.

With our estimated SD in hand, we can use it to compute the estimated SE.

**STEP** In cell H8 enter the label "estimated SE" and in cell H9 , enter the formula: *=H7/SQRT(3)*. Press F9 a few times.

There is a chain of reasoning, like one domino hitting another, and it goes like this: the estimated SE bounces because the sample SD bounces and it bounces because the sample SD is based on bouncing residuals. They bounce because y bounces and y bounces because it contains measurement error. The source of all the bouncing is measurement error.

The estimation strategy is to use the residuals as estimates of the unobservable errors. But does this work? Is the estimated SE a good estimator of the exact SE?

**STEP** Run a simulation with 10,000 repetitions of the estimated SE. How does it perform?

The results are quite disappointing. It is obvious that the estimated SE is biased. The average of 10,000 estimated SEs is around 0.51 and we know the exact SE is 0.577.

Why is the estimated SE performing so badly? Since n is fixed, there can be only one suspect: the estimated SD.

*STEP* Run a simulation with 10,000 repetitions of the estimated SD. How does it perform?

Again, we get simulation results that provide strong evidence of bias. Sigma is 1 and we are getting an average of 10,000 sigma-hats that is more than 10% under 1.

Can we fix this? Yes. Our estimators are performing badly because they are biased in small samples, but consistent. This means that as n rises, they do better.

*STEP* Return to the *DGP* sheet and copy it. Rename the sheet "DGP10" and modify it so that it implements the DGP with 10 measurements.

Be sure to start from the beginning and walk through the each of the cells, reminding yourself of what each cell is computing or displaying.

*STEP* Run a simulation with 10,000 repetitions of the estimated SE of the sample average. How does it perform?

Much better, but it is still biased. We can use our rough "margin of error" computation on the sim results to show that the average of 10,000 estimated SEs is too far away from $\frac{1}{\sqrt{10}}$. We cannot use "simulation bounce" to explain away this deviation from what we expected.

*STEP* In cell J4 of your MCSim sheet, enter the formula: *=J6/100.*

This shows that the precision of the simulation is such that we have evidence of bias. On the other hand, it is much better than n=3. As a percentage deviation from the Exact SE, the bias in the Estimated SE has fallen from over 10% with n=3 to around 3% with n=10.

To conclusively show that the estimated SE is a consistent estimator (we have already shown it is biased in small samples), we can run one last sim.

*STEP* Copy the *DGP10* sheet and rename it "DGP100." Take 100 measurements and simulate the estimated SE. What happens?

Your simulation results show that the sample average estimator is extremely close to the exact SE with n=100. We can conclude that estimating the exact SE with $\hat{sigma}$ works quite well for large sample sizes in the Measurement Error DGP.

Our work has shown that with 3 measurements we get a lot of bias in the estimated SE of the sample average. When we made 10 measurements, the amount of bias fell, but we could still see it with 10,000 repetitions. Finally, with 100 measurements, we cannot see the extremely small bias with our simulation—this offers a concrete, clear demonstration of the concept of consistency.

We close with an analogy that might help you understand and remember the concept of consistency: some infinite series converge and some do not. For example, this geometric series gets closer and closer to 2 as you process more terms:

$$\sum_{n=0}^{\infty} \left(\tfrac{1}{2}\right)^n = 1 + \tfrac{1}{2} + \tfrac{1}{4} + \tfrac{1}{8} + \cdots = 2$$

On the other hand, this harmonic series diverges:

$$\sum_{n=1}^{\infty} \frac{1}{n} = 1 + \frac{1}{2} + \frac{1}{3} + \frac{1}{4} + \cdots$$

The more terms you add, the bigger it gets. Of course, you can use a spreadsheet to confirm that the geometric series gets closer and closer to 2 as you add more terms, while the harmonic series keeps getting bigger and bigger.

Consistency is, like an infinite series, a concept about what happens asymptotically, as you approach a limit, but never actually reach it. Just like convergence or divergence depends on the specific series, consistency or inconsistency depends on the specific statistic being used and the DGP.

We want our estimator to be unbiased, but if not, we want it to be consistent because that means the bias will go to zero as the number of observations increases. If an estimator is biased and inconsistent, that means the bias does not go away as n rises. This is bad.

## 15.4 A Road Map

This is the second time we have computed and discussed the three SEs: Approximate, Estimated, and Exact. There are many things going on at once. Perhaps a broad overview will help us put everything in its proper place.

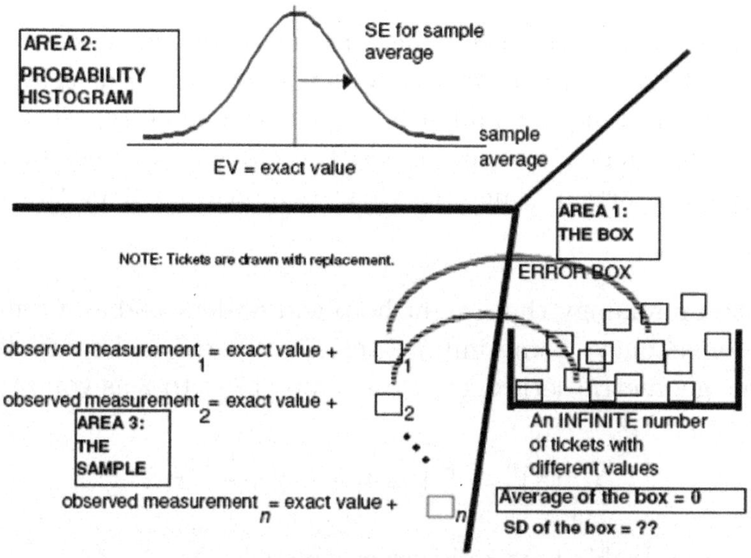


Figure 15.2: A Road Map for the Measurement Error Model

Begin in Area 1 in Figure 15.2. The error box has an infinity of tickets, but the average of all of them is zero if the measuring process is unbiased. So errors are positive and others negative, some are big and others small, but they average to zero. The SD of the errors is usually unknown and this is why we do not know the Exact SE. We have to compute the Estimated SE with an estimate of sigma (the SD of the errors).

Next, look at Area 2 in Figure 15.2. This shows a histogram of all of the sample averages. We never see this in the real world. We only see one sample (which is Area 3). Monte Carlo simulation is giving us an approximation to that perfectly smooth, continuous histogram because you can not sim an infinity of repetitions.

Finally, consider Area 3. That is the one thing we can see in the measurement error DGP. Each time we measure, get an observation. It is the sum of an unseen exact value and random draw from the error box.

Here is a summary of the whole thing: with our one random sample, we compute the sample average. It is our estimate of the exact value. We know that our sample average is likely to be off the exact value by its Exact SE. We cannot compute the Exact SE because we do not know $\sigma$ (the SD of the errors), so we estimate it by plugging in an estimate of the SD of the errors in the formula, $\frac{\hat{\sigma}}{\sqrt{n}}$.

## 15.5 Takeaways

Measurement error was the first DGP for inferential statistics.

The story is simple, yet it took real intellectual breakthroughs to figure out what was going on. In essence, measurement is tainted by noise that cannot be separated from the truth.

We can lower the volume on the noisiness by averaging the measurements.

The sample average is a random statistic, so it has a sampling distribution.

The Polling and Measurement Error DGPs share the same formula to estimate the exact SE, but they tell different stories about the process that generated the data.

You have to know the DGP that generated your data. You cannot just take data and apply formulas. If the data do not obey the DGP, the formulas are inappropriate and will mislead you.

The estimated SE of the sample average is a biased estimator of the exact SE of the sample average because the estimated SD ($\hat{\sigma}$) is a biased estimator of $sigma$.

Consistency says that as $n$ increases, the estimator's bias falls. This is a desirable property.

We would prefer, of course, an estimator that was unbiased, so that no matter the sample size it was centered on the truth, but a consistent estimator is the next best thing.

The introduction said that Gauss and others showed the best way to use the data—compute the sample average estimator. Explaining the superiority of the sample average estimator remains on our agenda.

Perhaps the biggest takeaway is that when we observe something, we are actually seeing a mix of truth and noise. This means that some numbers contain randomness and, therefore, they bounce around.

It does not have to be measurement error that makes a number bounce. Remember father's and son's heights? The idea is that a son's height has a fixed component that depends on the father's height, but also a random portion from the DNA sequence each son happens to get.

Therefore, we can interpret height as a random variable, but not because of measurement error. If you measure again tomorrow, you will get the same height. But the genetic lottery at play means that if you imagined rewinding and replaying the world so the same son was born again, he would have a different height (from DNA randomness).

Really understanding that a number might not be a pure, constant amount and explicitly explaining the source of its bounciness is a huge step in statistical mastery. It is a difficult idea, but we will keep working on it.

# 16. Regression with Measurement Error

The measurement error data generation process can be applied to a regression setting. Instead of estimating a single parameter, $\mu$, we estimate a relationship between $x$ and $y$ with an intercept, $\beta_0$, and slope, $\beta_1$. We will focus on the slope.

Our slope estimate contains random errors from a measurement process so the slope estimate is itself a random variable. The estimator has a sampling distribution with expected value and standard error.

As usual, there are three SEs: Exact, Approximate, and Estimated. The second row of LINEST displays Estimated SEs and we will explain how they are computed.

Although our data are hypothetical, for context, we will base our example on *Hooke's Law.* This says that the extension of a spring when weight is attached is a linear function of the weight, as shown in Figure 16.1.

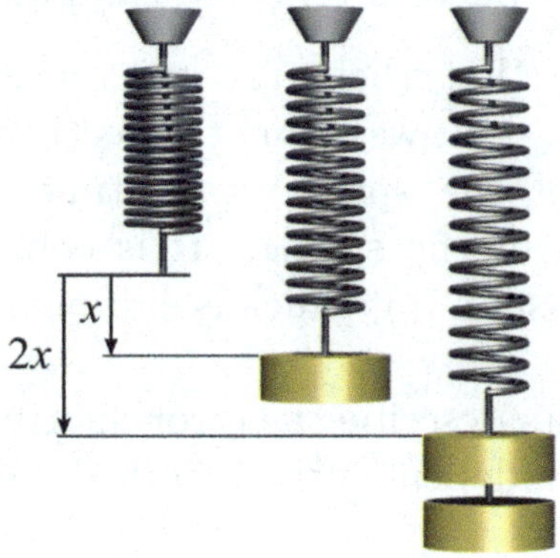


Figure 16.1: Hooke's Law.
Source: en.wikipedia.org/wiki/Hooke%27s_law

The length ($y$) of the spring without any weight ($x$) is $y = \beta_0$. The length with the weight depends linearly on the amount of the weight, $y = \beta_0 + \beta_1 x$. "Beta-zero" is the intercept and "beta-one" is the slope.

Every spring has its own constant of proportionality, $\beta_1$, the slope in the equation of the line, depending on its own springiness. A small $\beta_1$ means the spring is really tight (like the coil in car's suspension system) and barely stretches when you add weight.

We recognize, of course, that Hooke's Law has limits. It fails completely if you apply such a large weight that you deform the spring.

## 16.1 Implementing the Measurement Error DGP

The Measurement Error DGP in a regression setting is an extension of the measurement error model for the sample average. Instead of adding measurement error to a single parameter, $\mu$, we add it to a linear equation.

Formally, the DGP of the Measurement Error regression model is:

$$y_i = \beta_0 + \beta_1 x_i + \epsilon_i, i = 1, ..., n$$

$$E[\epsilon_i] = 0, i = 1, ..., n$$

$$SD[\epsilon_i] = \sigma, i = 1, ..., n$$

$$\epsilon_i \text{ is distributed independently of } \epsilon_j$$

$$x_i \text{ fixed in repeated sampling}$$

The first equation says that there are true, exact values for the intercept, $\beta_0$, and slope, $\beta_1$, but we cannot directly observe them. Instead, we see $y$ (the length of the spring) each time we apply a weight ($x_i$) and measure the length of the spring. It is composed of the true, exact length ($\beta_0 + \beta_1 x_i$) plus epsilon ($\epsilon$), which is a draw from an error distribution.

The next three statements describe the error distribution. The measurement errors are unbiased and independently, identically distributed, *iid.*

The last requirement means that the weights applied to the spring are exactly the same throughout the measuring process. Say we have five weights from 100 to 500 grams. We will use these exact weights as we measure the length of the spring. If we get an entirely new set of measurements (equivalent to taking another sample), we use the same set of weights.

*STEP* Name a sheet *DGP*. In cells A1 and A2, enter the values 10.12345 and 0.12345. Name these cells *beta_0* and *beta_1*. In cells B1 and B2 enter the labels *intercept* and *slope*. In cell A3 enter the value 4 (which is unrealistically high, but used for illustration), name the cell *sigma*, and in cell B3, enter the label *sigma*. Select cell range A1:B3 and make the text color red.

These are the parameters in the model. We would not know these in the real world. Our focus is on estimating $\beta_1$, the constant of proportionality of a particular spring.

The true, exact length of the spring with no weight on it is 10.12345 centimeters (about 4 inches). As weight is added, the slope, in units of cm/g, tells us how much the spring stretches. For every gram of weight, its length increases by 0.12345 cm.

*STEP* In cell A5 enter the label $x$ and in the five cells below enter the series 100 to 500 by 100. The biggest weight is 0.5 kg (just over a pound). In cell B5 enter the label *true* $y$ and in cell B6, enter the formula: *=beta_0+beta_1*A6*. Fill it down. Make it red text.

This is the true, exact relationship between the length of the spring and weight. We cannot directly observe it. Instead, we observe lengths that have measurement error baked into them.

*STEP* In cell C5 enter the label *epsilon* and in cell C6, enter the formula: *=NORMALRANDOM(0,sigma)*. Fill it down. Make it red text.

This is the stochastic or random component in the DGP. We add it to true $y$ to produce observed $y$.

*STEP* In cell D5 enter the text $y$ and in cell D6 enter the formula: *=B6+C6*. Fill it down. Create a chart of length as a function of weight. Press F9 a few times.

The dots are bouncing because $y$ is bouncing because the epsilons are bouncing. Neither the weights, nor true $y$, nor the parameters bounce.

*STEP* Click on a dot on the chart, then add a trendline and display the equation on the chart, along with $R^2$.

The regression describes the data well. The dots are near the fitted line so $R^2$ is close to 1.

This is all fine, but we are interested in estimating $\beta_1$. The equation of the fitted line gives us a *point estimate*, but we want to know much more about the sampling distribution of $b_1$.

The symbols really matter here. $\beta_1$ is the true, exact, unknown, constant parameter and $b_1$ is a single estimate that is a random variable with a sampling distribution. We want to know its center, spread, and shape.

## 16.2 The Sampling Distribution of $b_1$

We begin our analysis of the slope estimate by adding the true relationship to the chart. This offers another opportunity to cement the idea that the fitted line is an estimator with built-in variability.

*STEP* Click on a dot, copy the SERIES formula, press the *esc* key, click on the top-left corner of the chart so the formula bar is clear, click in the formula bar, and paste. Change the y cell range from column D to B so that the formula is: *=SERIES(Sheet1!$B$5,Sheet1!$A$6:$A$10,Sheet1!$B$6:$B$10,1)*. Change its *Chart Type* to straight line without markers and make the line red. Double-click the $y$ axis and set the min to 0 and max to 90.

Your result should look similar to Figure 16.2. The fitted, dashed line is pretty close to the true, exact relationship given by the red line.

| | |
|---|---|
| 10.12345 | intercept |
| 0.12345 | slope |
| 4 | sigma |

| x | true y | epsilon | y |
|---|---|---|---|
| 100 | 22.46845 | 1.826256 | 24.29471 |
| 200 | 34.81345 | -5.01295 | 29.8005 |
| 300 | 47.15845 | -0.21566 | 46.94279 |
| 400 | 59.50345 | 1.538437 | 61.04189 |
| 500 | 71.84845 | 4.545258 | 76.39371 |

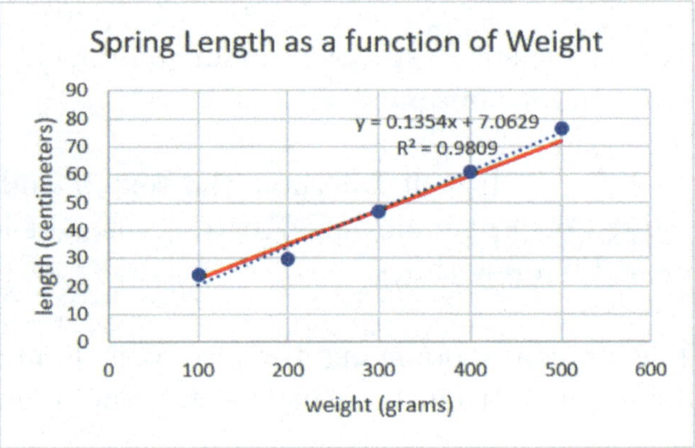


Figure 16.2: True line fixed while fitted line bounces.

*STEP* Press F9 repeatedly.

It is easy to see in Excel that the fitted line is dancing around the true, exact line—which does not move at all. This is crucial. Each sample of measurements produces a set of dots to which we fit a line. But since the measurements contain noise, so does the fitted line.

How noisy are the estimates? The intercept and the slope have their own sampling distributions, but we really only care about the slope so we focus on it.

We need a way to get the coefficients of the fitted line on the spreadsheet so we can use the MCSim add-in. LINEST is perfectly suited for this.

STEP Select cell range F6:G10 and enter a LINEST formula that regresses $y$ on $x$: *=LINEST(D6:D10,A6:A10,1,1)*. Remember to press ctrl-shift-enter since we are using an array function. You will know it is correct since the coefficients from the top row of LINEST must equal the fitted line equation in the chart. Press F9 a few times.

The coefficients and every other cell displayed by LINEST except G9 are bouncing. Cell G9 is reporting the degrees of freedom for the regression, which is the number of observations minus the number of parameters estimated, $n - k = 3$.

From previous work, we know the bottom row is reporting the regression sum of squares, RSS, and sum of squared residuals, SSR. The third row is reporting $R^2$ $(1 - \frac{SSR}{RSS+SSR})$ and the root mean squared error, RMSE. These are all random variables now because every time we take a new set of five measurements, we get new $y$ values.

Since our focus is on estimating the spring's constant of proportionality, $\beta_1$, we use the MCSim add-in to repeatedly remeasure, compute $b_1$, and directly observe the results.

Instead of calling $b_1$ the slope estimator, we introduce its more formal name, Ordinary Least Squares, OLS. Ordinary is an adjective used to distinguish this recipe from an almost countless number of alternatives. Ordinary is the basic, simplest recipe: choose intercept and slope to minimize the SSR.

STEP Use the MCSim add-in to track cell F6 with 10,000 repetitions.

Your results show that the average of 10,000 estimated slopes is quite close to $\beta_1 = 0.12345$ which is evidence that the OLS slope estimator is unbiased. The approximate SE is about 0.0126. This is the amount of variability we can expect to see in a slope from one set of five measurements.

Is the exact SE of $b_1$ equal to $\frac{\sigma}{\sqrt{n}}$ just like the exact SE of the sample average? No, but it is closely related. The exact SE of the OLS slope estimator is

$$\frac{\sigma}{\sqrt{n}SD_x}$$

The new $SD_x$ term in the denominator reflects the influence of the $x$ variable, the weights in this case, on the bounce of the SE of the slope. Below we show that the bigger the $SD_x$, the smaller the SE of $b_1$.

STEP In cell L6 of your simulation results sheet, enter the formula: *=sigma/(SQRT(5)*STDEV.P(DGP!A6:A10))*.

Notice that we use the population SD of the $x$ variable. This is because the $x_i$ are fixed in repeated sampling. They do not contain sampling error and we are not using them to estimate anything. Notice also how closely the simulation's SD approximates the exact SE.

What about the third SE—where is the estimated SE?

STEP Return to the *DGP* sheet and look at the second row of LINEST. These are the estimated SEs for the slope and intercept.

The intercept has its own formula, but we can compute the estimated SE of $b_1$ to replicate LINEST's output. The equation is

$$\frac{\hat{\sigma}}{\sqrt{n}SD_x} = \frac{RMSE}{\sqrt{n}SD_x}$$

Just like with the sample average, where use the sample SD to estimate $\sigma$, we use the RMSE, which is the SD of the residuals adjusted for degrees of freedom, to estimate $\sigma$.

STEP In cell E7 enter the formula: *=G8/(SQRT(5)*STDEV.P(A6:A10))*.

There is nothing mysterious in regression output. Everything has a recipe. Knowing how statistics are computed helps you understand what they mean and how they work.

But one outstanding question remains: How good is the estimated SE of $b_1$ at estimating its exact SE? There's a sim for that.

*STEP* Simulate 10,000 repetitions of cell F7. Compare the average of the sim to the exact SE. What do you find?

It may not look like it, but your results show a problem. Instead of announcing that the numbers are close enough since they are so small, we can compute the percentage deviation from the truth.

*STEP* In cell L5 of simulation results sheet, enter the formula: *=(J5-MCSim2!L6)/MCSim2!L6*.

You will find a roughly 8% point difference. That is a lot. In fact, the estimated SE of $b_1$ is a biased estimator of its exact SE. Why?

*STEP* Return to the *DGP* sheet and simulate 10,000 repetitions of cell G8, the RMSE. Compare the average of the sim to $\sigma = 4$. What do you find?

If you compute the percentage deviation of the average of 10,000 RMSEs to the truth, $\sigma = 4$, you will see that the RMSE is about 8% off.

The estimated SE of $b_1$ is a biased estimator of the exact SE because the RMSE is a biased estimator of $\sigma$. Then why do we use it?

As with the SE of the sample average, the bias gets smaller and smaller as the number of observations rises. This is the property of *consistency* in an estimator. An unbiased estimator is unbiased for any $n$, but a consistent estimator needs a "large" sample size to perform well.

*STEP* Copy the *DGP* sheet and extend the DGP to take 10 measurements. In your new sheet, copy cell range A6:D10, select cell A11, and paste. Modify the chart, LINEST, and the formula in cell E7 so that all of the data are used. Run a simulation of cells F7 and G8. What do you find?

The percentage deviation of the estimated SE of $b_1$ goes down to around 3% and the average of 10,000 RMSEs is much closer to 4.

Doubling the number of observations again, to 20, with four observations with each weight drives the bias down to about 1%. This demonstrates that the estimated SE of $b_1$ is a consistent estimator of its exact SE.

## 16.3 The Spread of the Xs

The formula for the SE (exact or estimated) makes clear that it depends on $\sigma$ (or an estimate of $\sigma$), $n$, and the $SD_x$. The first two are easy to understand: (1) more noise in measurement produces a bigger SE and (2) more observations leads to a smaller SE (following the square root of $n$). The last factor, $SD_x$, deserves some attention.

*STEP* Copy the DGP sheet and rename it "DGPSDx." Change cell A10 from 500 to 1,000. Change the max value on the chart to 160 so it displays properly. Press F9 a few times. What effect does this have on $b_1$?

The chart shows that the last weight is now much farther away from the center so the line is much longer. There is still bounce in the fitted line, but it does not move around as much as it did before. This is evidence that the SE of the slope has fallen.

As a purely mechanical fact, we have greatly increased $SD_x$ by changing the last weight to 1,000 grams.

*STEP* Use STDEV.P to compute the population $SD_x$ in cell A12 of the *DGP* and *DGPSDx* sheets.

The standard deviation of the weights has more than doubled, from about 141 g to over 300 g. What effect does this have on the SE($b_1$)? The formula tells us, but we can confirm with simulation.

*STEP* In cell I6 of the *DGP* sheet, enter the formula: *=DGPSDx!F6.* Above it, enter the label "Slope from DGPSDx sheet." Access the MCSim add-in and track both slopes, in cells F6 and I6, making sure to check the *Recalculate the Entire Workbook* option.

Your results should be similar to Figure 16.3. This offers conclusive and visual proof that the sampling distribution of $b_1$ is more concentrated around $\beta_1$ the bigger the spread of $x$.

| SDx 141 | | SDx 316 | | |
|---|---|---|---|---|
| $F$6 | | DGP!$I$6 | | Notes |
| Average | 0.123 | Average | 0.123 | Both unbiased, but |
| SD | 0.0129 | SD | 0.0057 | SE much smaller for |
| Max | 0.174 | Max | 0.146 | SDx = 316. |
| Min | 0.074 | Min | 0.099 | |

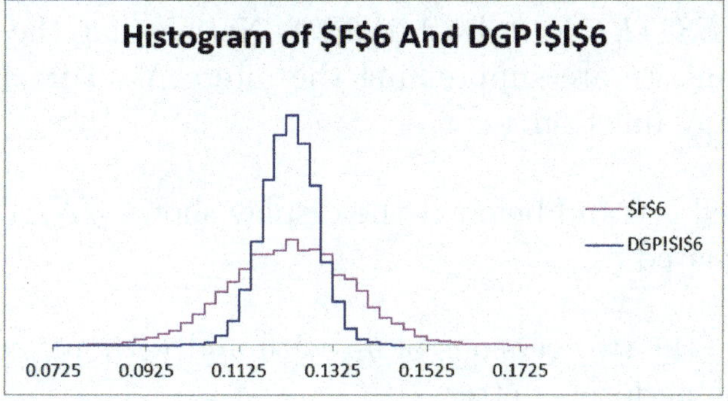


Figure 16.3: The bigger $SD_X$, the smaller SE($b_1$).

We do not usually get to control the $SD_x$, but if we did, we would certainly prefer to extend $x$ over a broader range because it would stabilize the bounce in the fitted line. This is equivalent to saying that we get to enjoy a smaller SE.

## 16.4 Regression without an Intercept Term

What if we knew the intercept with certainty? Suppose the manufacturer has exactly made a spring to our specifications so that we know it is 10.12345 cm long.

Now, we do not have to estimate the intercept and our DGP is:

$$y_i = 10.12345 + \beta_1 x_i + \epsilon_i$$

We can write this as:

$$y_i - 10.12345 = \beta_1 x_i + \epsilon_i$$

This is not merely an algebraic restatement. The rewritten version is said to have no intercept. Or, we could view it as having an intercept of zero. But how would we run a regression with this DGP?

*STEP* Return to the *DGP* sheet. In cell K5 enter the label "y-intercept" and in cell K6 enter the formula: *=D6-beta_0*. Fill it down.

We can now use the regression without a constant term option in LINEST. This is what "no intercept" means.

STEP Select cells M6:N10, click in the formula bar, and enter the formula: *=LINEST(K6:K10,A6:A10,0,1)*. Notice that the third argument is 0, which means we are suppressing the intercept. Hit *ctrl-shift-enter* to compute the array function.

The top-right cell is 0 and below it the display shows *#N/A* to indicate that there is no estimated SE.

We can compare the two versions of $b_1$, with and without an intercept term, to see which one performs better.

STEP Run a simulation of cells F6 and M6 with 10,000 repetitions.

Your results show the version without the intercept is superior. They are both unbiased estimators of $\beta_1$, but the approximate SE (the SD of your sim) is more than halved without the constant term. We want both accuracy (unbiased) and precision (low SE) so we would use the no intercept regression to estimate the constant of proportionality in our spring.

This result is common sense. By incorporating more information, such as knowing that $\beta_0 = 10.12345$, we improve our estimate of $\beta_1$. The challenge is often how to incorporate relevant information into the regression.

## 16.5 Takeaways

Bivariate regression with a Measurement Error DGP is an extension of the model with a single parameter, $\mu$.

One important addition is that $x_i$ are fixed in repeated sampling. If the X variables contain noise (from measurement or elsewhere), the DGP is different and we cannot trust the usual regression output.

The exact SE of the slope in a bivariate regression is an extension of the formula for the exact SE of the sample average:

$$\frac{\sigma}{\sqrt{n}SD_x}$$

And, just as with the sample average, in the real-world, we estimate the unknown $\sigma$ with its sample analogue, RMSE. This produces the estimated SE.

The estimated SE version of the exact SE formula is what the second row of LINEST is computing for the slope. The intercept has a different formula.

Yet another similarity is that the estimated SE of $b_1$ is a biased, but consistent estimator of the exact SE.

The connections between the SEs of the average and slope should not be surprising since we know regression is in the same family as the average.

The formula shows that the SE depends on three factors, $\sigma$, $n$, and $SD_x$. We showed that, if we could choose the $x_i$, we would make them be far apart.

If we know the true, exact value of the intercept, we can run a regression without an intercept to improve our estimate of the slope.

There are no confirmed drawings or paintings of Hooke. He was reclusive and fought bitterly with Newton, who was not nice to Hooke.

Figure 16.4: Robert Hooke (1635-1703) generated by ChatGPT.

# 17. MultiVar Reg with Measurement Error

The Measurement Error DGP in a multivariate regression is the same as the bivariate version except the equation for $y_i$ has more than one $x$ variable.

Formally, the DGP of a multivariate Measurement Error model is:

$$y_i = \beta_0 + \beta_1 x_{1i} + \beta_2 x_{2i} + ... + \beta_k x_{ki} + \epsilon_i, i = 1, ..., n$$

$$E[\epsilon_i] = 0, i = 1, ..., n$$

$$SD[\epsilon_i] = \sigma, i = 1, ..., n$$

$$\epsilon_i \text{ is distributed independently of } \epsilon_j$$

$$x_{ki} \text{ fixed in repeated sampling}$$

There are $k$ independent variables influencing $y$. We will estimate $k + 1$ parameters because each $x$ variable has a slope parameter, $\beta_k$ plus the intercept, $\beta_0$. Remember that the estimates are $b_k$.

The DGP says nothing about the connection of the $x$ variables to each other. We will see that the higher the correlation, the higher the SEs of the slopes. There is no specific value, but high correlation between x variables is called *multicollinearity* and if it equals 1.0, it is called *perfect multicollinearity.*

*STEP* In a blank Excel workbook, name a sheet "DGP." In cells A1, A2, and A3, enter the values 100, 2, and 8. In cells B1, B2, and B3 enter the labels "beta0," "beta1," and "beta2." In cell A4 enter the value 5 and add the label "sigma" in cell B4. Select cell range A1:B4, click the *Formulas* menu item in the Ribbon and click *Create Names from Formulas.* Check the "right column" and click OK. This names cells A1:A4 with the labels in column B. Select cell range A1:B4 and make the text color red.

Our focus is on estimating $\beta_1$, but as you will see, the $x_2$ variable also matters even if we care only about the effect of $x_1$ on $y$.

To create the independent variables, we use the BIVARNORMAL(mean x, SD x, mean y, SD y, r) array function. It works just like NORMALRANDOM, but it draws two random variables (like throwing two dice) at once with given means and SDs and imposes the given $r$, correlation coefficient. These functions live in the *MCSim* add-in so install it now if it is not already installed.

*STEP* In cells A5 and B5 enter the labels "x1" and "x2." As with all array functions, we first select cell range A6:B6, then click in the formula bar, and enter the formula: *=BIVARNORMAL(20, 5, 100, 0.1,0.9)*. Press ctrl-shift-enter.

Two numbers appear on the screen. You can see that $x_1$ is much smaller than $x_2$ which makes sense since $x_1$ has mean 20, while $x_2$ has mean 100.

*STEP* Press F9 a few times and watch the numbers.

Clearly, there is not much bounce in $x_2$ since its SD is 0.1. On the other hand, $x_1$ is producing a much bigger range, with numbers from 10 to 30 or so.

*STEP* Select cell range A6:B6 fill this down to row 15. In cell B17 enter the formula: *=CORREL(A6:A15,B6:B15)*. Press F9 a few times.

Again, it is easy to see that the two $x$ variables are following the means and SDs you entered in the function. The correlation coefficient parameter, however, requires more explanation.

*STEP* Create a scatterplot of $x_2$ as a function of $x_1$. Press F9 a few times.

The cloud of ten points is often cigar-shaped because $r = 0.9$ is a pretty strong association. Recall that $-1 \leq r \leq +1$. Zero is a formless blob, with no linear association, and the cloud gets tighter as $r$ approaches 1 in absolute value. The sign controls the slope (up or down) of the line around which the dots are clumping.

*STEP* Run a simulation of r to see that the BIVARNORMAL function function is working as advertised.

Your sim results show that the average of the correlation coefficients is smaller than 0.9. This is due to the small sample size. Since we are about to deaden the cells, this does not matter.

*STEP* Copy the *DGP* sheet and rename it "DGPFixedX." Press F9 a few times and when r is close to 0.9 (it does not have to be exactly 0.9), copy cell range A6:B15 and paste values. Press F9 a few times.

The $x$ variables are no longer bouncing. This meets the requirement of the DGP that they be fixed in repeated sampling.

*STEP* We add epsilon with the NORMALRANDOM(0, sigma) function in column C (make it red text) and produce $y$ in column D with the formula *=beta0+beta1*A6+beta2*B6+C6.*

We are ready to regress $y$ on the two $x$ variables. We cannot use the trendline approach because that only works for bivariate regression. Instead, we simply use LINEST to get the estimated coefficients.

*STEP* Select cell range F6:H10, click in the formula bar, and enter the formula: *=LINEST(D6:D15,A6:B15,1,1).* Press ctrl-shift-enter. Add labels in reverse order across the top, "b2," "b1," and "b0."

We have implemented the DGP and we have a way to regress $y$ on $x_1$ and $x_2$ for each sample. We are ready to explore the sampling distribution of $b_1$.

## 17.1 The Three SEs Again

There are always three SEs for sample statistics, but their analytical formulas differ depending on the DGP and recipe of the statistic.

The exact SE is always the true, correct SE. The approximate SE is from a simulation and, if the DGP is properly implemented, it can be made to be as close to the exact SE as desired by running more repetitions. The estimated SE is the one we usually have and it can be untrustworthy. It assumes that the DGP requirements have been met. If they have not, it is wrong.

For the Measurement Error DGP, we know that the sample average has an exact SE given by the formula:

$$\frac{\sigma}{\sqrt{n}}$$

With measurement error in a bivariate regression, we know the exact SE includes the same basic formula, but we add the $SD_x$:

$$\frac{\sigma}{\sqrt{n}SD_x}$$

Is this the exact SE of the Measurement Error DGP with two x variables that have $r \approx 0.9$? Let's find out.

*STEP* Run a simulation of $b_1$ with 10,000 repetitions.

We have evidence that $b_1$ is unbiased because the average of 10,000 $b_1$s is close to $\beta_1 = 2$. The SD of the 10,000 $b_1$s is an approximate SE. Is the exact SE the formula from the bivariate regression?

*STEP* In cells A18 and B18 of the *DGPFixedX* sheet enter the labels "SDx1" and "SDx2." In cell A19 enter the formula: *=STDEV.P(A6:A15)*. Fill it right. In cells F12 and G12 enter the labels, "exact SE b2" and "exact SE b1." In cell F13 enter the formula: *=sigma/(SQRT(10)*B19)*. In cell G13 enter the formula: *=sigma/(SQRT(10)*A19)*.

We can see if this formula for the exact SE works by comparing it to the approximate SE from the simulation.

*STEP* Return to the *MCSim* sheet and, in cell L6 enter the formula: *=DGPFixedX!G13*.

The approximate SE and our exact SE candidate are far apart. This tells us that the exact SE formula is different in the multivariate regression setting. There is a missing piece in the formula. We are not capturing the relationship between $x_1$ and $x_2$.

To compute the exact SE for $b_k$ with multiple $x$ variables, we must incorporate the connection among the $x$ variables.

$$\frac{\sigma}{\sqrt{n}SD_{x_k}\sqrt{1-R_k^2}}$$

The new term in the denominator is the R$^2$ from an *auxiliary regression* of $x_k$ on the other independent variables. Although it sounds fancy, as we know from earlier work, an auxiliary regression is a regression that is not of primary interest. Auxiliary means supplementary or additional help and that is exactly the role of an auxiliary regression.

There is some intuition in the R$^2$ term from the auxiliary regression. It is capturing how much information the independent variables share. The best case scenario is an R$^2$ of zero because then the other $x$ variables do not explain any of the variation in $x_k$. Thus, we are getting a completely separate channel of effect on $y$. With $R_k^2 = 0$, the square root of 1 is 1 and the term drops away.

Our two $x$ variables, however, are highly correlated by construction. Thus, they will produce a high $R^2$ and the square root of one minus $R^2$ will be small. Since the term is in the denominator, the smaller it is, the bigger is the exact SE.

We can put the formula for the exact SE of $b_1$ to the test and demonstrate how the auxiliary regression works in practice.

*STEP* Select cell range J6:K10, click in the formula bar, and enter the formula: *=LINEST(A6:A15,B6:B15,1,1)*. Press ctrl-shift-enter. Press F9 a few times.

The cells do not bounce because the "exes are fixed in repeated sampling." With the $R^2$ from the auxiliary regression, we can modify our exact SE of $b_1$.

But before we do, notice that the $R^2$ is quite high. That is because we constructed the two $x$ variables to have a high correlation. The square of $r$ is $R^2$.

*STEP* In cell J12 enter the formula: *=B17^2* and compare the result to the $R^2$ in the auxiliary regression.

This confirms the claim and may leave you wondering why we bothered to do the auxiliary regression. This is because with more than two independent variables, we cannot use the square of $r$ shortcut.

We return to the formula for the exact SE of $b_1$.

*STEP* Click in the formula bar of cell G13 and add the new term so the formula is now: *=sigma/(SQRT(10)*A19*SQRT(1-J8))*.

How does this exact SE compare to your sim results of $b_1$?

*STEP* Return to the *MCSim* sheet and compare the approximate and exact SEs.

This time they are quite close. This tells us that the exact SE formula is correct. If they are still far apart, you know there is something wrong in your computation of the exact SE.

Can we use our new formula for the SE to replicate the estimated SE produced by LINEST? Of course.

*STEP* In cells G15 enter the label "est SE b1." Enter the formula bar of cell G13, copy the entire formula, select cell G16, click in the formula bar, and paste. Replace "sigma" with "G8" and press enter. Press F9 a few times

We exactly replicated the estimated SE of $b_1$ in LINEST.

How well does the estimated SE of $b_1$ perform?

*STEP* Run a simulation of cell G7 with 10,000 repetitions.

Your results show that the average of 10,000 estimated SEs is close to the exact SE, but it is biased. We have seen this before. As the sample size increases, the bias gets smaller. This is consistency.

The fact that the SE of a sample slope rises as the included independent variables are more connected is called *multicollinearity.*

At its extreme, if $R_k^2 = 1$, the regression cannot be fit with the included $x$ variables. This is called perfect multicollinearity.

In practice, there is always some multicollinearity among the included independent variables. If two variables are strongly connected with each other, the SEs for their slopes will be high. The regression is telling us that it cannot disentangle the individual effects of the independent variables. We will see what can be done in this situation in future work.

## 17.2 Zero Correlation

We close with a demonstration of the claim made above that zero correlation between an independent variable and the other included $x$ variables lowers the variability of the estimated slope.

The easiest way to set the scene is by taking advantage of the work we have already done. We will create a set of $x$ variables that have almost zero correlation.

*STEP* Copy the *DGPFixedX* sheet and rename it *DGPFixedXr0*. Return to the DGP sheet and change the *Calculation* setting (in the *Formulas* tab on the *Ribbon*) to Manual. Change the correlation in all of the BIVARNORMAL functions from 0.9 to 0. Press F9 repeatedly until cell B17 is close to zero, within ±0.05. Copy cell range A6:B15, go to the *DGPFixedXr0* sheet, select cell A6, and paste values. Reset calculation to automatic.

The correlation is low, but we want it to be zero. We can do this with Solver.

*STEP* In the *Solver Dialog* box, make the objective function B17, click the *Value of* option, and make the changing cell B15. Click Solve.

Cell B15 changes slightly and the correlation coefficient is extremely close to zero.

The auxiliary regression shows that $x_2$ has no explanatory power. Notice that the exact SE of $b_1$ is much smaller than before.

A comparison of the *DGPFixedX* and *DGPFixedXr0* sheets does not hold everything constant except $r$. The $x$ values are different so the $SD_x$ values are different.

It is, however, true that by creating $x_1$ and $x_2$ with zero correlation, we have improved the precision of the estimator of $\beta_1$.

*STEP* Confirm that calculation is automatic and if not, make it so.

Manual calculation is dangerous. It is an Excel, not a workbook, setting. When on, Excel will not recalculate formulas as they are entered. This can make it seem like Excel is broken or, worse, display a cell value that is not correct because it has not been processed. Be careful when using this feature.

## 17.3 Takeaways

In a multivariate regression setting, the Measurement Error DGP requires another extension to the formula for the exact SE of a sample slope:

$$\frac{\sigma}{\sqrt{n}SD_{x_k}\sqrt{1-R_k^2}}$$

The $R_k^2$ term is capturing the relationship between the independent variables. As $R_k^2$ approaches 1, the SE is inflated. This is called multicollinearity.

An auxiliary regression is a supplementary regression. We ran an auxiliary regression to find the $R_k^2$ term.

LINEST, of any other regression software, does not compute the estimated SE with an auxiliary regression. It uses matrix algebra routines.

The formula for the exact SE makes clear how regression is connected to the sample average in that it uses the same sigma over the square root of $n$ with additional terms.

# 18. The Classical Econometric Model

Measurement error is the dominant data generation process in the natural sciences, but not in the social sciences.

We introduce the Classical Econometric Model (CEM) with a hypothetical example designed to highlight essential features of the DGP.

The goal is to create an intuitive, engaging story that will stay with you. It is based on the real-world sport of skiing, but we construct an absurd scenario that will serve to make it memorable.

## 18.1 Olympic Skiing

The women's super giant slalom at the 1998 Winter Olympics in Nagano, Japan, was incredibly close. The Gold Medal went to Picabo (pronounced PEEK-a-boo) Street of the United States with a time of 78.02 seconds. In second place, with a time of 78.03 seconds, was Austrian skier Michaela Dorfmeister and another Austrian, Alexandra Meissnitzer came in third at 78.09 seconds.

In the 2018 Winter Olympics in South Korea, it happened again. Esther Ledecka of the Czech Republic won gold in a time of 81.11 seconds and Anna Veith of Austria won silver in 81.12 seconds. It is a different mountain and course so you cannot compare these times to the 1998 Nagano times.

In terms of the overall Olympic Super-G competition, there are about 40 skiers. The slowest skiers are usually around 5 to 10 seconds behind the winner. The course is about two kilometers and average speed is about 100 kmh (60 mph).

The Measurement Error DGP would say that each skier actually had the exact same time on her run in the Super G, but the official clock sometimes registered a faster time, sometimes a slower time. The winner is just the luckiest skier.

This is nonsense, but just to be clear, we can write down the DGP. All of the skiers in the competition, from 1 to $n$, have the same true time, but when we measure the time, we add measurement error.

$$ObservedTime_i = TrueTime + \epsilon_i$$

Notice that there is no subscript on *TrueTime*. This means it is the same for all of the skiers. The only reason why they get different times is because we introduce random noise when we measure the time of each skier.

Why doesn't the Measurement Error DGP apply to Olympic skiing results? The timing system uses photocells and is actually quite precise. There is measurement error, but not enough to explain the variation in times. Also, assuming that the true time is the same for all skiers means they are all equally good and this does not make common sense. Finally, even if they were exact clones, they would not have the same time on every run. Snow, wind, and the path left by other skiers must have an impact on each skier's time. Measurement error does not describe this data generation process so we should not use it to analyze the data.

We want a DGP that actually reflects how the skier's time is produced. We know that luck is involved, but it is not from measurement error. It is the randomness of shifting body weight at a specific moment or a gust of wind.

We know there are many factors that influence a skier's time, but we will focus on just two: training and talent. The first factor considers the effort put in through preparation, coaching, and practice. The second factor is innate—each skier has an inborn, genetically-determined, raw ability to move muscles and react as they fly down the mountain. Some people are just great athletes and can do amazing physical feats of running and jumping.

This DGP can be written as a formal equation:

$$ObservedTime_i = \beta_0 + \beta_1 Training_i + \beta_2 Talent_i + \nu_i$$

The error term, the Greek letter nu ($\nu$) is not measurement error. Nu represents the inherent randomness in the world. If the same skier, with the same training and talent, took another run, the time would be different because of many small effects of wind and movement as they went down the mountain. We interpret a negative $\nu$ as good luck on a particular run.

We can add that $\nu$ is well-behaved: it has mean zero, constant spread, and is iid.

So, this is the CEM? No. There is a crucial step remaining before we get to the CEM.

There is an important difference between training and talent: one is easy to see and the other is not. We can directly observe how many hours per day an athlete trains, but there is no way to directly measure talent.

Since we cannot quantify talent, we cannot estimate the parameters in the DGP that produces observed ski times. We have to work with a different version of the DGP that includes talent in the error term.

Formally, we can write the DGP as:

$$ObservedTime_i = \beta_0 + \beta_1 Training_i + \epsilon_i$$

$$\epsilon_i = \beta_2 Talent_i + \nu_i$$

The source of the error term in the CEM can be found in two places. First, each error term contains omitted variables (an independent variable that influences the dependent variable, but is not included in the regression). We have stressed talent (natural ability) as an obvious cause of performance, but there are many more omitted variables such as the motivation and health of each skier. The second component of the error in each observation is $\nu$, the inherent randomness in the world (slight wind shifts while flying down the mountain, bumps, etc.) or, in other words, just plain luck.

The error term is $\epsilon$ just like in the measurement error DGPs we have studied, but the explanation is completely different. There is measurement error, but it is so small as to not really matter. Instead, $\epsilon$ is the sum of all omitted variables (like talent) and random chance on a particular run down the mountain.

The fact that omitted variables are in the error term raises worries about omitted variable bias or confounding. We will see, concretely and visually, that the omitted variables that influence $y$ must be independent of any included $x$ variables.

## 18.2 The Austrian Experiment

Suppose the Austrian Ski Federation, responding to national outcry over the Olympic failures of their skiers decides to figure out the effect of training and talent on ski times.

Everyone knows that training can improve performance in just about every aspect of life, but exactly how much of an effect it has is unclear. The same is true for talent. Obviously, you have to be a great athlete to be an Olympic skier, but how much would time improve if you increased talent? Figuring out how much training and talent affect performance is really hard to do.

The Austrian Ski Federation knows that they cannot measure talent, but they decide to focus on the effect of training. They want to estimate $\beta_1$, how more training impacts time.

They embark on an outlandish plan. They will kidnap people, force them to train an assigned number of hours each day, and then have them compete in a race. They will use the race results and training time records to estimate the effect of training.

We will implement this experiment in Excel. Figure 18.1 has the parameters and results for one race. With $\beta_1 = -0.5$, each additional hour per day of training, holding talent constant, lowers time 0.5 seconds. This is the parameter the Austrian Ski Federation wants to know.

| | A | B | C | D | E |
|---|---|---|---|---|---|
| 1 | 80 | beta0 | (secs) | | |
| 2 | -0.5 | beta1 | (secs/hours per day) | | |
| 3 | -0.2 | beta2 | (secs/index points) | | |
| 4 | 1 | sigma | (secs) | | |
| 5 | | | | | |
| 6 | Skier | Training (hours per day) | Talent (index points) | nu (secs) | Time (secs) |
| 7 | A | 0 | 2.102390547 | 0.390663 | 79.97 |
| 8 | B | 1 | 1.01369511 | -1.13997 | 78.16 |
| 9 | C | 2 | -5.91367545 | 0.440695 | 80.62 |
| 10 | D | 3 | 0.34414384 | 0.083246 | 78.51 |
| 11 | E | 4 | 8.509781167 | 1.111645 | 77.41 |
| 12 | F | 5 | 8.543180314 | 0.21368 | 76.01 |
| 13 | G | 6 | 2.896027962 | -0.18726 | 76.23 |
| 14 | H | 7 | -5.20672071 | 1.757416 | 79.3 |
| 15 | I | 8 | -0.5878318 | -0.35341 | 75.76 |
| 16 | J | 9 | -0.5808887 | 0.005689 | 75.62 |

Figure 18.1: Setup and results for one race.

Talent is measured as an index, with zero being average and the bigger the number, the greater the person's natural ability. Skiers E and F in Figure 18.1, have the most talent; skiers C and H are not natural-born skiers.

With $\beta_2 = -0.2$, each index point increase in talent, holding training constant, lowers time by 0.2 seconds.

Figure 18.1 shows skier J won the close race, with a time of 75.62, 0.14 seconds faster than skier I. Neither of them are especially talented, but they were forced to train a lot.

Remember, the Austrian Ski Federation kidnaps people and *randomly* assigns each of them a training regimen. It is critical for the success of the experiment that it be random assignment.

*STEP* Create your own version of Figure 18.1. Name the sheet *DGP* and enter the parameters and values in columns A and B. Name the parameter cells. For Talent, enter this formula: *=NORMALRANDOM(0,4)*. For nu, enter: *=NORMALRANDOM(0,sigma)*. For Time, enter this formula: *=ROUND(beta0+beta1*B7+beta2*C7+D7,2)*. Press F9 a few times.

*Training* is fixed in repeated sampling, but *Talent* is bouncing. Each new set of 10 skiers have their own natural ability and they are randomly assigned to a number of hours per day of training. Skier A always does no training, hanging out in the chalet, while skier J is forced to practice 9 hours per day.

*STEP* In cell C18, compute the correlation coefficient between training and talent by entering the formula: *=CORREL(B7:B16,C7:C16)*. Press F9 a few times.

The correlation coefficient is bouncing. What is its expected value?

*STEP* Run a simulation of $r$.

Although a single sample will have a nonzero correlation, over many samples, there is no correlation between training and talent. This is because the Austrian Ski Federation randomly assigns the skiers.

*STEP* In cell G6 enter the label "Winner" and in cell G7, enter the formula: *=MIN(E7:E16)*.

This is the fastest time, but which skier won? The MATCH function returns the position of a value in a given range. It has three arguments: MATCH(lookup value, lookup array, optional match type).

*STEP* In cell G8 enter the formula: *=MATCH(G7,E7:E16,0)*. Press F9 a few times.

Excel displays the number of the winning skier. This number is a random variable produced with sample data so it has a sampling distribution.

*STEP* Run a simulation of cell G8.

As expected, skier J wins the most often, followed by I, and so forth. Skier A almost never wins because the people assigned to that get no training at all.

*STEP* In cell range I7:J11, regress time on training by entering this formula: *=LINEST(E7:E16,B7:B16,1,1)*. Press F9 a few times.

The results look promising. Cell I7, which is $b_1$ our sample estimate of $\beta_1$, is around $-0.5$.

*STEP* Run a simulation of cell I7.

Your results are evidence that we have an unbiased estimator of $\beta_1$ because the average of the slopes is close to the parameter value. The SD is an approximate SE that is reliable as long as the simulation is properly implemented. We will get estimates of $\beta_1$ that are $-0.5 \pm 0.14$.

*STEP* Return to the *DGP* sheet and sim the estimated SE of $b_1$.

The average of the estimated SEs is reasonably close to the approximate SE in the simulation of $b_1$. All is well.

Notice that we do not need the exact SE. With advanced math we could find the formula and compute it, but the simulation's approximate SE provides us with a reliable substitute for the exact SE.

We could improve our estimator in two ways. We could increase the number of Austrians we force to train or we could space out the training more, say from 0 to 18 hours a day (although we might exhaust our subjects if we do this—one can train too much). We know from previous work that the $n$ and $SD_x$ are in the denominator of the analytical formula of the SE so increasing them will lower the SE.

Unfortunately, someone at the Austrian Ski Federation has come up with an idea that will not improve our estimator. In fact, it will break it.

## 18.3 Independence of the Omitted Variables

"After we kidnap them, we should not randomly assign, but give more athletic people more training so we can create Olympic champions!"

How can we implement this idea in Excel so we can see what it does to our estimator of $\beta_1$? We will use the RANK function. It returns the rank of a list of numbers given three arguments: RANK(number, lookup list, optional order).

*STEP* Copy the *DGP* sheet. In cell F6 enter the label "Rank" and, in cell F7, enter the formula: *=RANK(C7,$C$7:$C$16).* We do not need the third argument because our order is smallest to largest and this is the default. Fill it down. Press F9 a few times.

You can see that the RANK function is working as advertised. The smallest number (most negative) is 10 and the biggest number is 1.

*STEP* Click on cell B7 and enter the formula: *=10-F7.* Fill it down.

We have implemented the idea to give more training to those with more talent. (We conveniently assume that we can observe talent.) What will this do?

The spreadsheet is flashing red flags. The correlation coefficient between training and talent is quite high (the plan does this) and, most importantly, $b_1$ seems quite far from $-0.5$. We need to run a simulation to see how connecting training and talent has affected the sampling distribution of $b_1$.

*STEP* Run a simulation of cell I7.

The results are devastating. Our slope estimator is now biased. It appears that one more hour of training per day gives an 0.7 or 0.8 decrease in time when we know it is really 0.5 seconds. We broke the slope estimator when we made the omitted variable, talent, which is included in the error term, correlated with training. This induced omitted variable bias in our slope estimator.

It looks like training is more important than it actually is because part of the improvement in performance is that more highly trained people are also more talented. Training is getting too much credit.

It is easy to be confused so we want to make sure that the role of the omitted variables is clear. The CEM has an error term that relies heavily on variables that influence $y$, but are not included in the regression. The omitted variables in the CEM error term are often difficult to measure, like talent. They become part of the noise of the DGP.

As long as these omitted variables are independent, not connected, with included $x$ variables, the regression estimators of the slopes are unbiased.

If, however, there are omitted variables that are connected with the included $x$ variables, then we get omitted variable bias in our slope estimators.

For example, economists are constantly running earnings function regressions, with some measure of earnings explained by education, experience, and demographic characteristics. But what about intelligence, motivation, and ambition? If these omitted variables, which influence earnings, are also correlated with education, our estimate of the rate of return to education is compromised. It will seem like education is more responsible for higher earnings than it actually is because education is capturing the channel through which the omitted variables affect earnings.

At this point, most students think, "OK, now tell me about the omitted variable bias test." There is no such thing. There is no formula or diagnostic that reveals omitted variable bias.

There is an Omitted Variable Bias Rule, which we used in our earlier work on multivariate regression. If you recall, it said this:

$$g_1 = b_1 + d_1 b_2$$

The Omitted Variable Bias equation says that the slope on training in the misspecified regression that omits talent ($g_1$) is biased by an amount equal to the slope of the auxiliary regression of talent on training ($d_1$) times the slope of talent in the correct model ($b_2$).

The Omitted Variable Bias Rule is useless in helping us identify actual omitted variable bias in a real-world setting. After all, you would have to know $b_2$ and if you did, then why would you run the misspecified model in the first place?

The value of the Omitted Variable Bias Rule lies in formally connecting how an omitted X variable will cause trouble and makes clear that there are only two ways out: the omitted X is independent of the included X ($d_1 = 0$) or the omitted X does not affect Y ($b_2 = 0$). Neither is expected to hold in any real-world application.

Statistical analysis is art and science. The mechanics of regression, the formulas for the SEs, and so forth have a strong mathematical foundation. But evaluating and judging the specification of a regression model, thinking about which variables are missing that could bias the slopes of included $x$ variables, is truly an art. It requires repeated practice and exposure to many different regressions.

## 18.4 A Formal Version of the CEM

A multivariate version of the Classical Econometric Model can be written like this:

$$
\begin{gathered}
y_i = \beta_0 + \beta_1 x_{1i} + \beta_2 x_{2i} + ... + \beta_k x_{ki} + \epsilon_i, i = 1, ..., n \\
\epsilon_i = \beta_{k+1} x_{k+1,i} + ... + \beta_{k+j} x_{k+j,i} + \nu_i, j = 1, ..., m \\
E[\nu_i] = 0, i = 1, ..., n \\
SD[\nu_i] = \sigma, i = 1, ..., n \\
\nu_i \text{ is distributed independently of } \nu_j \\
x_i \text{ fixed in repeated sampling} \\
\text{omitted } x_{k+j,i} \text{ independent of fixed included } x_{ki}
\end{gathered}
$$

A crucial part of the CEM's DGP is the second equation. It says that the error term, $\epsilon$ is composed of perhaps many omitted $x$ variables plus a traditional disturbance term, $\nu$ (with the usual iid properties).

The last statement reminds us that the omitted $x$ variables cannot be connected with the included $x$ variables. If so, we get omitted variable bias in our estimated $b_k$ slopes.

Almost all empirical papers in economics will include the first equation, but they often fail to discuss explicitly the sources of error.

## 18.5 Takeaways

Austria has snow-capped mountains, the Alps, where alpine skiing originated. Austria has won more alpine skiing Olympic medals than any other country.

The Austrian Ski Federation, however, has never kidnapped people and forced them to train in an effort to determine the efficacy of training.

The example and story, even though fictional, is a good way to understand the Classical Econometric Model because the idea that training and talent are two causal factors affecting performance is true in many areas of our lives.

The CEM, sometimes called the Classical Linear Model, has an $\epsilon$ that does not rely on measurement error. Instead, a great deal of the noise in $y$ is because of omitted variables.

As long as these omitted variables are independent of the included $x$ variables, everything is fine and the slope estimators of the included $x$ variables are unbiased.

The moment independence is broken, so is the estimator. It becomes biased, confounded by the effect of the omitted variable.

Random assignment protects against omitted variable bias because it ensures that independence is maintained.

We broke the estimator by forcing highly talented people to get more training, but we also could have broken the independence requirement by simply *allowing* people to choose their level of training.

Since people tend to do more of what they are good at, letting them decide what to do results in omitted variable bias. This is why random assignment is superior to observational studies.

Of course, random assignment is not easy. How would it work for education? We randomly assign people to different levels of schooling (forcing some to work right out of high school and others to go to graduate school) and then watch how much they earn over their lifetimes. That is even more absurd than Austrians forcing random people to ski an assigned number of hours per day!

# 19. The Gauss-Markov Theorem Univariate

The Gauss-Markov Theorem (GM) says that if the data generation process (DGP) meets certain requirements, then the ordinary least squares (OLS) regression estimator is the best linear unbiased estimator (BLUE).

The short way to state GM is: *OLS is BLUE.*

The short way ignores the "if" part, but it remains true that the Gauss-Markov Theorem applies *only* if the DGP is well-behaved and the sample data we have actually meet the requirements.

The goal is to understand the Gauss-Markov Theorem by demonstrating it with simulation. We will implement DGPs in Excel and race alternative estimators against each other.

There is little mystery here because the Gauss-Markov Theorem says OLS will win, but setting up and seeing the races makes clear what the Gauss-Markov Theorem is all about and why we use OLS.

We proceed in two steps:

1. Univariate: we estimate the average of a list of numbers, $\mu$, from sample data
2. Bivariate: we estimate the slope in a bivariate relationship, $\beta_1$, from sample data

The theorem is named after Gauss and Markov because Gauss proved it with DGP requirements that included normality of the errors. Then, a century later, Markov, working on stochastic processes in Russia, proved it with less restrictive requirements.

Proving Gauss-Markov is a staple of advanced statistics courses. Our simulation work can only demonstrate the claim. Simulation alone can never prove anything.

As you work through the DGPs and simulations, remember the key point: there are many estimators (recipes) to choose from and we want to use the best one.

## 19.1 The Best Univariate Estimator

Figure 1 lays out the landscape. Start at the top left and read clockwise. The circle is all possible estimators. We can choose any estimator. We reduce the field of candidates by eliminating those that do not meet certain conditions.

The vertical ellipse is the subset of estimators that are unbiased. We rule out those that are not in the ellipse because they are not centered on the truth. These estimators are systematically wrong, like a dart throwing style that produces outcomes not concentrated on the bullseye.

The horizontal ellipse is the subset of estimators that are linear. There are other theorems that do not limit the competition to only linear estimators, but Gauss-Markov imposes this requirement, which is explained below.

The intersection of the two ellipses contains the estimators that are both linear and unbiased. The Gauss-Markov Theorem is a statement about only the estimators that are both linear and unbiased.

In addition to unbiasedness, another desirable property is minimum variance. The winner is the estimator with the smallest SE (the square root of the variance). We want the most precise estimator.

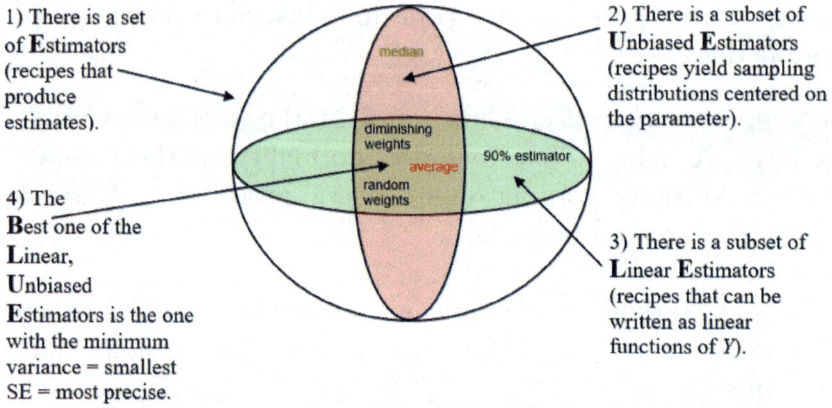


Figure 19.1: A map of the Gauss-Markov universe.

The average, diminishing weights, and random weights in Figure 19.1 are examples of linear and unbiased estimators. The median is an example of unbiased estimator, but it is not linear. The 0.9 estimator is linear, but biased. The Gauss-Markov Theorem says nothing about the median or 90% estimators.

Linear estimators are those that can be written as linear functions of $y$:

$$w_1y_1 + w_2y_2 + ... + w_ny_n$$

The $w_i$ are called *weights*. For the sample average, the weights are $\frac{1}{n}$.

The diminishing weights estimator has different weights, that get smaller and smaller, so it is a different linear estimator:

$$w_i = \frac{1}{2^i}, i = 1, ..., n-1 \text{ and } w_n = \frac{1}{2^{n-1}}$$

The random weights estimator draws weights randomly from a uniform distribution on the interval [0,1] (i.e., Excel's RAND function) except for the last weight which is:

$$w_n = 1 - \sum_{1}^{n-1} w_i$$

The 90% estimator uses weights that are 90% of the sample average weights.

The median is not a linear estimator because it cannot be written as a linear function of $y$. The median's recipe says to sort the $y$ values and pick the middle one (with special rules for even numbers of observations or repeated values of y).

We can demonstrate the Gauss-Markov Theorem in the univariate case with a Measurement Error DGP:

$$y_i = \mu + \epsilon_i, \text{ where } \epsilon_i \sim N(0, \sigma), iid$$

*STEP* In cell A1 enter the number 10 and in cell B1, enter the label "mu." In cell A2 enter the number 3 and in cell B2, enter the label "sigma." Name the two cells mu and sigma. In cell A4 enter the label "Obs" and create a series from 1 to 5 below it. In cell B4 enter the label "epsilon" and below it enter a formula that implements the DGP.

With the MCSim add-in installed, the NORMALRANDOM function is available. You could also use Excel's native function, NORM.INV with an embedded RAND: *=NORM.INV(RAND(),0,sigma).*

STEP In cell C4 enter the label $y$ and below it enter a formula that generates a data point according to the DGP. Fill it down. Press F9 a few times.

Your formula in cell C5 needs to use the DGP equation: *=mu+B5*.

To estimate $\mu$, we use the sample average. We do so because of the Gauss-Markov Theorem. To demonstrate how it works, we implement the four estimators above. We begin with the sample average.

STEP In cell D4 enter the label "Average" and in cell D5, enter the formula: =AVERAGE(C5:C9).

Instead of Excel's AVERAGE function, we could apply the linear estimator formula, using the weights for the average, $\frac{1}{n} = 0.2$.

STEP In cell E4 enter the label "w" and enter 0.2 below it. Fill down the 0.2 values. In cell E10 enter the formula: *=SUM(E5:E9)*. In cell F4 enter the label "wy" and in cell F5 enter the formula: *=E5*C5*. Fill it down to F9. In cell F10 sum the five $w_i y_i$ values.

The sum of the weights is always 1 and the sum of the $w_i y_i$ is always the same as Excel's AVERAGE function. This shows that the linear estimator formula for the average is working properly.

Our first competitor steps up: the diminishing weights estimator. We enter its weights, compute $w_i y_i$, and then sum to bring this estimator to life.

STEP Copy cells E4:F4, select cell G4, and paste. In cell G5 enter the formula: =1/2^A5. Fill down to cell G8. In cell G9 enter the formula: *=G8*. In cell G10 sum the weights. In cell H5 enter the formula: *=G5*C5*. Fill it down to H9. In cell H10 sum the five $w_i y_i$ values. Press F9 a few times.

Both estimators have weights that sum to 1, but they are producing different estimates of $\mu$ because they use different weights. Which one is better? We simulate to find out.

STEP Before we sim, name the cells to make the simulation results easier to understand. Name cell F10 "average" and H10 "diminishing." Use the MCSim add-in to run a simulation that tracks both cells F10 and H10.

Your results provide evidence that both are unbiased, but the average is more precise. Perhaps more repetitions, however, would provide more support.

*STEP* Run the same simulation, but with 10,000 repetitions.

Yes, now it is more clear. The sampling distribution of the average is definitely more spiked than its rival. This is desirable because the more spiked, the more likely we will draw an estimate close to the truth.

We can compute the exact SE for the sample average (it is $3/\sqrt{5} \approx 1.34$), but we do not know the formula for the exact SE of the diminishing weights estimator. This does not matter because we know the SD of 10,000 estimates is a good approximation to the exact SE.

Next up, the random weights estimator.

*STEP* Implement the random weights estimator (based on its definition above) in columns I and J. Race it against the average. What do you find?

Random weights is unbiased, but its SE is huge compared to the sample average. We much prefer the sample average to the random weights estimator.

We could keep going, inventing estimators with different weights than the sample average. As long as the weights sum to 1, the estimator is unbiased. Gauss-Markov says that no estimator in this class can have a lower SE than the sample average. This is why we use the sample average to estimate $\mu$.

Our work hints at the proof strategy used on this problem. It revolves around choosing weights to minimize the SE, subject to a constraint that the weights sum to 1. It can be shown that, for our DGP, The optimal weights are $1/n$, the sample average weights.

The 90% estimator is a biased estimator, but we can race it against the average to see how it compares.

*STEP* Implement the 90% estimator in columns K and L. Based on its definition above, its weights are 0.18 for each observation. Race it against the average. What do you find?

Look carefully at the sim results. The 90% estimator is biased (did you notice its weights did not sum to 1?) so it is not in the competition, but it has a smaller SE than the average. Even though the darts are more tightly clustered, they are not centered on the truth.

You might wonder if the trade-off could be worth it so that we would prefer a slightly biased estimator that is quite precise. That is a difficult question beyond our scope and it is outside the bounds of Gauss-Markov—only unbiased estimators are allowed in the competition!

We close our univariate competition with one final race—average versus median. The median, like the average, is a common measure of central tendency. We know the median is not a linear estimator so Gauss-Markov cannot tell us anything about the outcome of this race. Simulation, however, can tell us if the median is a better choice than the average.

*STEP* Enter the formula for the median of $y$ in cell F11: *=MEDIAN(C5:C9)*. Run a sim of cells F10 and F11. What do you find?

The median is unbiased, but it has a bigger SE so our champion, the sample average, remains undefeated. The Gauss-Markov Theorem guarantees that the average cannot be beat by any rival—we cannot come up with a linear unbiased estimator with a lower SE. This is why we use the sample average as our estimator of $\mu$.

In earlier work, we saw that the sample average is the least squares estimator: it minimizes the SSR. We use the sample average and not another estimator because of the Gauss-Markov Theorem.

## 19.2 Takeaways

The Gauss-Markov Theorem is foundational. It asks and answers a critical question: given some data, what is the best recipe to use? The answer is the sample average.

One important idea is that we are aware of the possibility that another statistic might beat the average using a single sample of data. This is the same as a cake tasting better one time.

We want the best recipe—for many, many cakes. The cake analogy says there is an ideal, perfect cake. There are many possible recipes to make it. Of all the recipes that usually hit the ideal cake, we want the one with the smallest dispersion.

Sure, any one time another recipe might beat our winner, but over many bakings, the winning recipe will win the most. That is why we bake the cake with that recipe.

We did not prove the Gauss-Markov Theorem. That requires analytical work with formulas. That was Markov's contribution.

Figure 19.2: Andrey Markov (1856-1922) generated by ChatGPT.

And just in case, to make sure there is no confusion, a *theorem* in mathematics is different from a *theory*. A theorem is a statement that has been proven to be absolutely true.

A theorem is like a fact that will never change. Unlike a theory that can be revised if we find new evidence, a mathematical theorem is true forever.

# 20. The Gauss-Markov Theorem Bivariate

The previous chapter showed that the sample average is BLUE for a DGP that adds noise to a constant. You cannot find a linear, unbiased recipe for the sample data that produces estimates of $\mu$ with a smaller dispersion than the sample average.

## 20.1 The Best Bivariate Estimator

We can do the same thing with regression that we did with the sample average. We cannot prove, but we can demonstrate with simulation that line fitting via OLS defeats a linear, unbiased alternative.

The Gauss-Markov Theorem says something much more encompassing than our simulation demo—there is *no possible estimator* that could beat OLS. It is guaranteed to be the minimum variance estimator in the subset of linear and unbiased estimators.

Figure 20.1 offers a visual to help us understand the situation. There are an infinity of rival estimators. In the class of linear and unbiased estimators (in the middle of Figure 20.1), we want the best one, defined as the one that is the most precise (has the minimum SE).

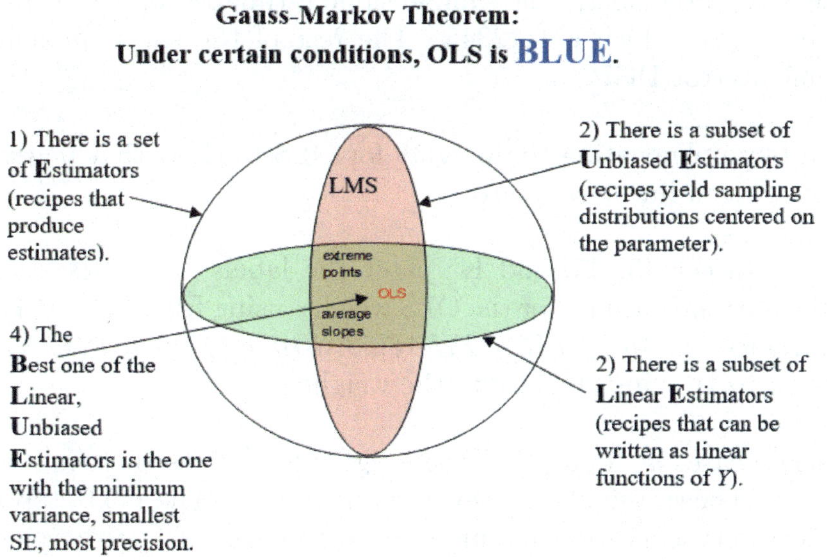


Figure 20.1: A map of the Gauss-Markov universe of estimators.

As we did for the sample average, we invent a few linear and unbiased rivals, extreme points and average slopes, then use simulation to race them against each other. LMS, least median of squares, is offered as an example of an advanced topic, robust regression.

We begin by showing that the OLS slope estimator can be written as linear functions of $y$:

$$w_1y_1 + w_2y_2 + ... + w_ny_n$$

For OLS, the weights are:

$$\frac{x_i - avg_x}{n \times var_x}$$

Again, we use a Measurement Error DGP, but this would also work with the Classical Econometric Model:

$$y_i = \beta_0 + \beta_1 X_i + \epsilon_i, \text{ where } \epsilon_i \sim N(0, \sigma), iid$$

*STEP* We save set up time by opening the measurement error with bivariate regression workbook (Hooke's Law) and saving as *GM.xlsx.* Next, delete all of the sheets except for the original *DGP* sheet. You can select multiple sheets by holding down the shift key as you click on the sheet tabs. Delete any cells to the right of LINEST in columns F and G. Finally, make sure the fifth observation's X value is 500.

Recall that cell F6 displays the sample slope estimator, $b_1$, of the population parameter, $beta_1$, which is 0.12345. The rest of the sheet implements the measurement error DGP.

We begin by implementing the weights for OLS to show that regression can be computed as a linear estimator.

*STEP* In cell I5, J5, and K5 enter the labels "w," "wx," and "wy." In cell I6, enter the formula for the OLS weight, using Excel's VAR.P function for the variance: *=(A6-AVERAGE($A$6:$A$10))/(5*VAR.P($A$6:$A$10))*. Fill it down to I10. In cell I11 add the weights.

In univariate case, we found that the weights for unbiased estimators had to sum to one. These weights, however, sum to zero. The next step reveals a second necessary condition for unbiased estimators in a regression setting (which are calling the bivariate case—we are estimating an intercept and a slope).

*STEP* In cell J6, multiply the weight by $x$ by entering the formula: *=I6*A6*. Fill it down. In cell J11 add the $w_i x_i$.

The $w_i x_i$ sum to exactly one. Is that a coincidence? No, for the bivariate case, an unbiased estimator has $w_i$ that sum to zero and $w_i x_i$ that sum to one.

*STEP* In cell K6, multiply the weight by $y$ by entering the formula: *=I6*D6*. Fill it down. In cell K11 add the $w_i y_i$. Name cell K11 "b1wavg" to indicate that it computes $b_1$ via a weighted average approach.

The result might surprise you—computing the OLS slope estimator as a weighted average of $y$ is identical to LINEST's slope estimator. This confirms that OLS is a linear estimator.

We will implement two alternative estimators in Excel: the extreme points and average slopes estimators. We will compare them to each other and OLS. There is no mystery here. Gauss-Markov tells us OLS will win, but it says nothing about who comes in second.

The extreme points estimator is based on the logic that the middle points do not matter. We can estimate the slope, $\beta_1$, with just the first and last points. The implementation is easy.

*STEP* In cell M5 enter the label "extreme points" and in cell M6, enter the formula: *=(D10-D6)/(A10-A6)*.

To express extreme points as a linear estimator, we use these weights:

$$w_1 = -\frac{1}{x_n - x_1}$$

$$w_i = 0 \text{ for } i = 2, ..., n-1$$

$$w_n = \frac{1}{x_n - x_1}$$

Notice that the weights are zero for the middle observations, which means their values have no effect on the slope at all.

*STEP* Copy the three labels in cells I5:K5, select cell N5, and paste. Implement the weights in column N, then compute $w_i x_i$ in column O and $w_i y_i$ in column P. Sum each column in row 11. Name cell P11 "extreme."

The weights summing to zero and the sum of $w_i x_i = 1$ tell us that the extreme points estimator is unbiased. Cell P11 is equal to cell M6, thereby confirming that we have simply rewritten the estimator as a linear function. How well does our two-point estimator perform compared to OLS?

*STEP* Run a simulation with 10,000 repetitions that tracks the OLS and extreme points estimators of the slope.

Your results show evidence that both estimators are unbiased, but OLS is more precise because it has a lower SD of 10,000 slopes. We use the SD as an approximation to the exact SE of the sample slope. Because we implemented the DGP based on the DGP equation, we can trust our simulation results.

We could compute the exact SE of the OLS sample slope estimator because we know its formula, but we cannot compute the exact SE of the sample slope for the extreme points estimator. Make no mistake, its exact SE exists, we just don't know the formula for it.

Fortunately for us, we have simulation and it gives us a reliable approximation to exact SE (and expected value and shape of the sampling distribution) of any statistic as long as we faithfully reflect the DGP in the Excel workbook.

*STEP* In cell M6 of your sim results, compute the percentage amount that the extreme points approximate SE exceeds that of the OLS estimator by entering the formula: *=(J6-L6)/L6.*

Your simulation results show that the extreme points estimator of $\beta_1$ does remarkably well given that it is only using two points. The SE of extreme points is about 10% bigger than OLS—not a big difference—which is why the histograms are pretty close to each other.

The average slopes estimator is an alternative way to estimate $\beta_1$. Its logic is to compute the slope from one point to the next, then average the slopes.

*STEP* In cell R5 enter the label "average slopes" and two cells below (in cell R7), enter the formula: *=(D7-D6)/(A7-A6).* This is the slope from observation 1 to 2. Fill it down to R10. In cell R11, compute the average of the four slopes and name cell R11 "avgslope."

The average slopes estimator can also be written with weights, but before we do anything else, you may have noticed that it is identical to the extreme weights estimator.

It turns out that if the $x$ values are evenly spaced, the two estimators are identical. We can make them different by changing one of the $x$ values.

*STEP* Change the $x$ value in cell A10 from 500 to 1000. Press F9 a few times and confirm that the two estimators are now different.

Next, instead of our usual approach where we rewrite the estimator as a sum of weighted terms, let's run a simulation to see how average slopes compares to the OLS slope estimator.

*STEP* Run a sim of cells K11 and R11. What happens?

One thing that jumps out is that the OLS slope estimator has improved. Since we increased the $SD_x$, the SE of the OLS sample slope has fallen.

Another clear result is that the OLS slope estimator easily defeats the average slopes estimator. The sampling distribution is much more concentrated and spiked for OLS compared to average slopes.

The average slopes estimator appears to be unbiased. If so, a weights-version of it would have weights that sum to zero and $w_i x_i$ that sum to one. In fact, this is correct.

Changing the last value of $x$ leads to a question. Presumably, in addition to improving the OLS estimator, this should improve the extreme points estimator. After all, the first and last points are now much farther apart so this might stabilize the estimated slope. Does it?

We can explore this question with simulation. We already know that, with $x_5 = 500$, OLS beats extreme points by about 10%. What happens to the two estimators as we increase the last value to 1,000 or 10,000?

Simulation shows that they both get better, as we expected and the difference gets smaller, but the percentage difference rises.

We close with a question that generated quite a bit of debate and controversy: does the distribution of the errors matter?

We can answer this question with simulation.

*STEP* Copy cell range A5:G10, select cell A14, and paste. Change the NORMALRANDOM function to UNIFORM in cell C15 and fill down. Run a simulation with 10,000 repetitions of cells F6 and F15. What do you conclude?

Your simulation results cannot be used to say that every error distribution with the same mean and SD produces the same sampling distribution, but it does provide evidence that there is no difference in the expected value and SE of the OLS slope estimator when using normally and uniformly distributed errors.

Careful inspection, however, does show there is a difference in the shape of the distribution. This difference disappears as $n$ rises.

## 20.2 Takeaways

There are an infinity of estimators. To estimate $\mu$ we could have used many different recipes for the sample data. There are even several variants of the average, such as the geometric and harmonic means.

Similarly, there are many other estimators for a slope in a regression equation. Even in the family of least squares, there are many alternative types of least squares, of which OLS is the most basic.

Figure 20.2 helps us understand how to evaluate estimators.

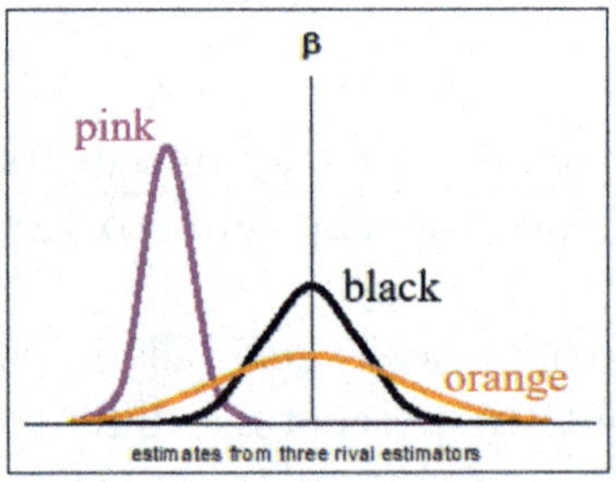


Figure 20.2: Judging estimators.

There is a parameter, $\beta$, for which we have three rival estimators. Pink is biased, so it is eliminated from consideration. Orange is less precise than black. With orange, on average, we get $\beta$, but we can also get estimates quite far from the truth. Of these three, the winner is black because it is unbiased and most precise.

The Gauss-Markov Theorem says that if the DGP is well-behaved, then OLS is BLUE. OLS is the *best* linear unbiased estimator for the well-behaved errors in the DGPs we have been working with. Best is defined as the most precise, and no rival in its class can beat it.

Well-behaved in the measurement and Classical Econometric Model DGPs means that the errors are centered on zero and iid—independently and identically distributed. In the CEM, we also need the omitted Xs that are in the error term to be independent of the included Xs.

The Gauss-Markov Theorem extends Figure 20.2 in the sense that it says that black is the winner of *all possible* linear, unbiased estimators.

The Gauss-Markov Theorem is why we use the "minimize the sum of squared residuals" recipe, which is known as OLS, to estimate the relationship between dependent ($y$) and independent ($x$) variables.

# 21. Robust Regression

The Gauss-Markov Theorem tells us OLS is BLUE, but in the real world there are many different estimators in use. Why is that? Because there is something about the DGP that makes OLS no longer the best estimator.

Our goal is to show an example of a situation where we would not use OLS, but opt instead for what is known as *robust regression*. Synonyms for the word robust include vigorous, sturdy, and stout. Robust regression emphasizes a regression algorithm's ability to withstand stress.

OLS performs poorly in terms of robustness because a single, aberrant data point, or *outlier*, can throw the fitted line way off.

It is easy to see an example of this with the average, which we know is closely related to OLS regression. The average of 1, 2, and 3 is 2. If 3 is changed to 30, the average shoots up to 11.

The median is a robust statistic because the median of 1, 2, and 3 is 2 and changing 3 to 30 does not change the median at all. The middle number is still 2. We say the median is robust to the change in a single number.

There is an analogue to the median in regression analysis called *least median of squares*, LMS. Instead of minimizing the sum of squared residuals, as OLS does, LMS minimizes the median of the squared residuals.

We will use simulation to compare OLS and LMS to understand how LMS performs and why it can be useful.

## 21.1 Dirty Data

OLS requires well-behaved error terms such as those provided by DGPs featuring measurement error or the Classical Econometric Model. But some DGPs feature dirty data, which include outliers or other extreme values.

Survey data often include answers that show that the respondent has misunderstood the question or otherwise do not make any sense. For example, in the CPS, HOURWAGE (the hourly wage in dollars per hour) has a few values of several cents per hour. We can delete these observations and go along our merry way, but what if the problem is more common?

Mount, et al. (1997) provide a nice real-world example of the problems associated with using least squares on dirty data. They were interested in computer vision.

The leftmost image in Figure 21.1 is a satellite photo. It is low resolution and heavily pixelated, but a human can easily see the road (the thick black line) with other objects scattered about.

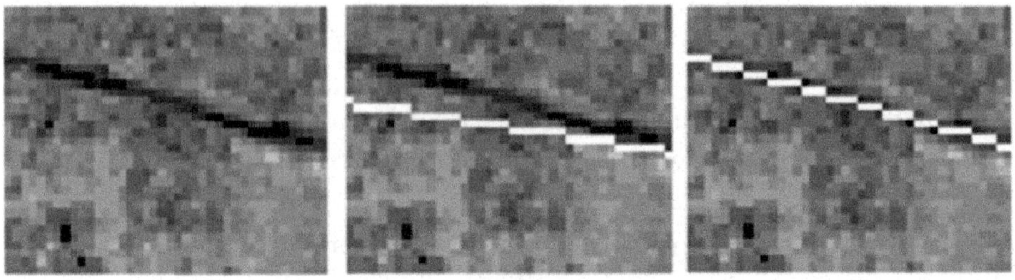

Figure 21.1: Finding the road in a satellite photo.

The middle photo shows an OLS fit to the data. Notice how the black points in the lower left-hand corner of the picture drag the least squares line down. The OLS fit misses the road.

On the right is a fit using LMS. It gets the road because it is robust to the noisy data. The points in the lower left (that might be a house) do not exert much pull on the fitted line.

Of course, OLS would perform well if we simply threw out the outliers (the points in the lower left-hand corner in Figure 1). This is exactly the strategy adopted by other robust estimators, such as *Least Trimmed Squares.* A rule is applied to detect outliers and they are trimmed, i.e., deleted, from the data.

Determining what constitutes an outlier, however, is not a simple task. Throwing out data points because they are inconvenient seems like a bad idea.

LMS, like OLS, uses all of the data, but uses a different recipe to fit the line. We turn to a concrete example in Excel to see how it works.

## 21.2 Implementing Dirty Data in Excel

We can use the measurement error DGP with bivariate regression to illustrate the idea of dirty data and explore how LMS works.

*STEP* Open the previous workbook on Gauss-Markov (*GM.xlsx*) and save as *LMS.xls*. Delete the sim sheets. In the *DGP* sheet, delete all columns to the right of column G and rows below 11, but keep the chart. If you do not have a chart, make a scatterplot of y on x, fit a trendline, and true y (as a thick red line). Finally, make sure that the fifth observation's X value is 500.

This implements the measurement error DGP with iid errors and the OLS estimated slope, $b_1$, is in cell F6.

*STEP* In cell C12 enter the number 100 and in cell D12, enter the label "outlier amount." Modify the last $\epsilon$ value by changing its formula to incorporate the outlier amount: *=NORMALRANDOM(0,sigma)+C12*.

The impact is dramatic. The fitted line in the chart has gone haywire. The OLS estimated slope is way off the true value of $\beta_1$. The last $y$ value is literally off the chart (you may have to unlock the $y$ axis to see it) and it has pulled the fitted line to it so the slope is much larger.

In a way, this result is not surprising. We no longer meet the requirements of the DGP because the last value of $\epsilon$ is not produced like the others. This violation has destroyed the ability of OLS to do a good job estimating the slope parameter.

What about LMS? How does it perform under these extreme conditions?

## 21.3 Least Median of Squares

OLS fits the line by choosing coefficients to minimize the sum of squared residuals, SSR. Formally, the optimization problem looks like this:

$$\min_{b_0,b_1} SSR = \sum_{i=1}^{n}(y_i - (b_0 + b_1 x_i))^2$$

There is an analytical solution, but we can show how it works with Excel's Solver, a numerical optimization tool.

To use Solver, we cannot have the numbers bouncing around. We have to work with a dead sample.

*STEP* Copy cell range A5:G10, select cell A14, and paste values. Press F9 a few times.

The numbers do not bounce so we know we have a single sample. We can use it to replicate LINEST's $b_1$ in cell F15.

*STEP* In cells J12 and J13 enter the labels "b0" and "b1." Enter a 1 in cells I12 and I13. Name those cells b0_ and b1_. In cell I14 enter the label "predicted y" and in cell I15, enter the formula: *=b0_+b1_*A15*. Fill it down to I19.

Next, we create columns for residuals and squared residuals, which we sum to give us our objective function.

*STEP* In cell J14 enter the label "residuals" and in cell J15, enter the formula: *=D15-I15*. Fill it down to J19. In cell K14 enter the label "squared residuals" and in cell K15, enter the formula: *=J15^2*. Fill it down to K19. In cell K20 enter the formula: *=SUM(K15:K19)*. In cell L20 enter the label "SSR."

We can now use Excel's Solver to find the SSR-minimizing coefficients, $b_0, b_1$.

*STEP* Click the *Data* tab, then click *Solver*. For the objective function, click cell K20, click *Min*, and for the changing cells, click cells I12 and I13. Confirm that "Make Unconstrained Variables Nonnegative" is unchecked. Click *Solve*.

Solver's $b_0$ and $b_1$ values are extremely close to those produced by LINEST.

Instead of the SSR, we could compute the average squared residual and minimize it.

*STEP* In cell L21 enter the label "avgSR" and in cell K21, enter the formula: *=AVERAGE(K15:K19)*. Run Solver. What happens?

The solution does not change. Minimizing the average of the squared residuals gives the same fit as when we used SSR.

This is not surprising since we are just dividing the SSR by $n$ so this has no effect on the optimal solution. But what if we did something else with the squared residuals, like take their median? This is LMS.

*STEP* In cell L22 enter the label "LMS" and in cell K22, enter the formula: *=MEDIAN(K15:K19)*. Run Solver. What happens?

LMS performs much better than OLS. LMS gives a slope estimate that is much closer to the $\beta_1$ than the OLS value around 0.3.

But how can we run a simulation with LMS? Excel does not have a LINEST-like function for LMS. Furthermore, we cannot find the LMS coefficients with Solver with bouncing numbers since every time Solver puts down a candidate solution, it recalculates the sheet.

## 21.4 LMS in Excel

While Excel does not have the ability to find the LMS fitted line, we can import a Visual Basic module with LMS code into our workbook. This provides access to an LMS function just like LINEST gives OLS results for a bivariate regression.

*STEP* Visit dub.sh/regcourse and download the *ModuleLMS.bas* file.

In Excel, we need to go to the Visual Basic editor, Excel's back end. From there, we can import the code in the downloaded file and then return to Excel and use it.

*STEP* From Excel, press alt-F11 to go to Visual Basic. Click *File*, then *Import File* (ctrl-m), and open the ModuleLMS.bas file. Press ctrl-q to close the editor and return to Excel.

The LMS code is open-source so you are free to read it, but this is not necessary. The LMS function is an array function and it requires a 3 x 1 cell range to display its output, $b_0, b_1$, and the median of the squared residuals. The LMS function accepts two arguments, LMS($y$ data cell range, $x$ data cell range).

We can test the LMS code on the dead sample to see how it compares to Excel's Solver.

STEP Select cell range N15:N17, enter the formula: *=LMS(D15:D19,A15:A19)*, and press ctrl-shift-enter.

Excel displays the LMS intercept and slope coefficients in the first two cells and the median of the squared residuals in the bottom cell.

For this problem, Excel's Solver often fails and it produces a solution for the minimum median of the squared residuals that is wrong. We can replace the LMS function solution in cells I12 and I13 to confirm that it is correct.

It turns out that the surface of the median of the squared residuals is quite different from their sum. Figure 21.2 shows the smooth surface of SSR with a well-defined minimum.

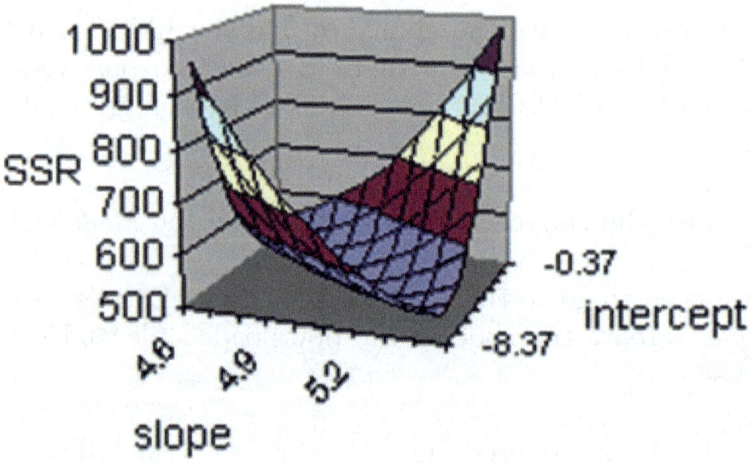


Figure 21.2: The smooth surface of the SSR.

Figure 21.3 offers two views of the surface of the median of squared residuals. It is like a crumpled piece of paper, with many local minima. This is a difficult problem for an optimization algorithm. It is easy for Solver to get stuck in what looks like a solution, but is not the global minimum.

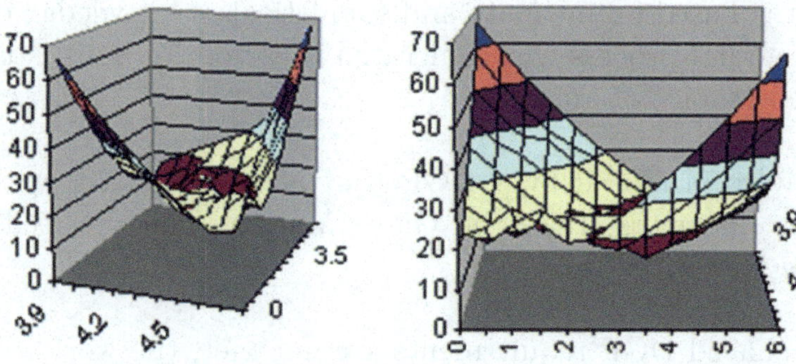


Figure 21.3: Two views of the jagged surface of LMS.

The LMS function uses a different approach to find the coefficients that give the smallest median of squared residuals (Barreto and Maharry, 2005). We can use it, just like LINEST, to provide LMS fits as the sheet recalculates and gives us new samples.

*STEP* Copy cell range N15:N17, select cell N6, and paste. Press F9 a few times.

Cells F6 and N7 are now reporting the OLS and LMS slope estimators. We can run a simulation to compare the performance of the two estimators.

*STEP* Use the MCSim add-in to track cells F6 and N7 with 10,000 repetitions.

The results are unambiguous: OLS is biased and LMS is not. The outlier value breaks OLS, but LMS is robust to the dirty data.

Then why don't we use LMS all the time?

*STEP* Change the outlier value to zero (returning the DGP to its well-behaved, iid condition) and simulate again.

This time, OLS beats LMS. Both are unbiased, but OLS has a much smaller SE so we would choose it over LMS.

You might expect the next sentence to describe the Dirty Data Test. Unfortunately, it does not exist. We do not have a mechanical way to decide whether or not to use OLS or LMS (or any other estimator, for that matter).

The decision is based on intimate and sophisticated knowledge of the DGP. What is the actual process that generated these data? This is the question that drives the choice of estimator.

If the data were produced by a DGP that is well-behaved, with a process that includes iid errors, then the Gauss-Markov Theorem tells us to use OLS because it is BLUE.

If the well-behaved DGP requirements are not met, the world of estimators is at our disposal and we have a difficult choice to make. Choosing the best estimator requires art (judgment) and science (mechanical knowledge of statistical properties of estimators). There is no rule to follow. The actual process that generated the data is the determining factor in deciding which estimator to use.

It is worth remembering that when you save an Excel file with a Visual Basic module in it, you must save it as an ".xlsm" file type. This filename extension means the file is a macro-enabled workbook. If you do not do this, the Visual Basic module is removed from the file.

Excel prompts you with a warning if you attempt to save a file with embedded Visual Basic code as a simple ".xlsx" file. If you want to keep the module with the workbook, click *No* when prompted to save as a macro-free workbook and then use *Save As* to save it as a ".xlsm" file type.

## 21.5 Takeaways

Robust regression is a category of regression analysis designed to withstand a variety of problems that plague data collection.

The idea is simple: like a bamboo tree that can withstand severe storms by bending, but never breaking, robust regression gives good results even in the presence of dirty data and outliers.

LMS is an example of a robust regression estimator. Instead of minimizing the sum (or average) of the squared residuals, the objective is to make the median of the squared residuals as small as possible.

The median is a robust statistic because it is impervious to extreme values. The average, on the other hand, is not. We say it has a low breakdown point—in fact, a single value can break the average.

To fit an LMS line, we imported a Visual Basic module into the workbook. It is good to know that the basic Excel environment can be extended with add-ins and modules that enhance Excel's capabilities.

> *EXCEL TIP* Visual Basic is the back-end of Excel where macros (code) are written and stored. A *.bas* filename extension indicates that the file can be imported into Visual Basic.

Knowing that LMS exists leads to a larger question: when should we use it? There is no easy answer to this. It depends on the DGP and the research question.

## References

Barreto, H. and Maharry, D. (2005) "Least median of squares and regression through the origin" *Computational Statistics and Data Analysis*, pp. 1391-1397, www.sciencedirect.com/science/article/abs/pii/S0167947305000046.

Mount, D,, Netanyahu, N., Romanik, K., Silverman, R., & Wu, A. (1997) "A practical approximation algorithm for the LMS Line Estimator," 8th Ann. ACM-SIAM Symposium on Discrete Algorithms, pp. 473-482.

Rousseeuw, P. and Leroy, A. (1987) *Robust Regression and Outlier Detection*, (New York: John Wiley & Sons).

# 22. Workshop

Set up your own DGP to see how all of this work that we have been doing ties together.

Use BIVARNORMAL to create two sets of two Xs in a regression of Y on X1, X2, X3, and X4. Use BIVARNORMAL for X1 and X2 and another BIVARNORMAL for X3 and X4. It doesn't matter what the mean and SD of each X is, but use a different mean and SD for each X variable. Make the first pair of Xs have $r = 0$ and you choose the correlation for X3 and X4. You decide how big a sample size to have.

Compute correlations of X1 and X2 and also for X3 and X4. Run simulations that show these are working as advertised.

Make up numbers for coefficients and sigma, then create epsilon and Y.

Organize the spreadsheet and make sure you clearly separate parameters from estimates.

Copy the sheet and rename it *DGPDead.* Deaden the Xs (you may have to set Calculation to Manual). Of course, the epsilons and Y are live and bouncing as you press F9.

Run a sim of $b_1$ to get the Approximate SE of $b_1$. In the sim results, write a note on how well the estimator did.

Do the same for another X (your choice).

In the *DGPDead* sheet, compute the estimated SE of $b_1$. You have to regress X1 on X2, X3, and X4. You will know you got it right if it agrees with LINEST.

Compute the Exact SE($b_1$).

Compare the Approx and Exact SEs. What do you conclude? Explain.

Run a sim of the Estimated SE($b_1$). How did it do?

Return to the *DGPDead* sheet and run a regression that is misspecified by dropping X4. What does this do to the regression?

# 23. SE for ± and Confidence Intervals

In the real world, we get a single random sample and use it to make estimates. With regression, we use English letters for coefficient estimates in the equation, $b_0, b_1, b_2$ and so on. These are estimates of population parameters, the Greek letters $\beta_0, \beta_1, \beta_2$, etc. This is called inference. We never actually know the value of the population parameters, the betas, but our estimates, the bees, are educated guesses of the beta values.

But estimates are only part of the story. Perhaps more important is the SE of the estimate. This tells us the variability in the estimate.

A further real-world complication is that we will not know the exact SE of each coefficient estimate. Sigma ($\sigma$), the SD of the error distribution, is one of the unknown parameters in the DGP. Thus, we have to rely on the estimated SE of each coefficient estimate, itself a random variable.

The usual convention is to drop the adjective *estimated.* This is confusing, but common. It means that if we say SE, we mean estimated SE. We always use the full name when we refer to the exact or approximate SE. We will adopt this convention to get you used to it —remember, "SE" by itself means "estimated SE."

The goal is to explain the three ways we can actually use an SE in real-world data analysis:

1. As a give or take, plus-minus number on the estimated parameter.
2. To create a confidence interval (CI) estimate for the parameter.
3. To conduct a hypothesis test of a claimed value of the parameter.

In the course of our demonstration of these three uses of the SE, we will use Excel's macro language, Visual Basic. We will write code that creates confidence intervals on the sheet.

Working on a concrete problem and writing code will help you with the extremely difficult language involved in correctly expressing the meaning of a confidence interval.

When we get to hypothesis testing, also known as tests of significance, be sure to be on the lookout for the common confusion between *statistical significance* and *practical importance.* These are nowhere close to the same thing. The ability to read and use words specifically and exactly will play a crucial role in mastering the concepts ahead.

## 23.1 Estimate ± SE

The easiest way to use an SE is to include it with the point estimate. If the regression slope coefficient is 23.45 and its SE 3.21, then simply report it like this, for example: "The estimated coefficient on $X_3$ is 23.45 ± 3.21."

This conveys the variability associated with the estimated coefficient. But let's step back and make sure we understand what is going on here.

Estimation implies we do not know the exact answer, the population parameter. Thus, we are well aware that our estimated coefficient is not a perfect measure of the parameter value: 23.45 is almost certainly *not* the exact answer. Instead, 23.45 is simply one estimate from its sampling distribution, like getting 9 on one roll of two dice.

By providing the SE, a knowledgeable reader can quickly get a sense of the estimated dispersion in the sampling distribution. The 3.21 number says that if we sampled many times, our estimates would have a bounce of about 3.21.

The ± SE number also signals what is not reasonable for the parameter value. It is possible for the parameter to be literally any number—after all, we do not know it. However, the estimate ± SE gives us a way to judge what is so unlikely that we don't believe it, based on the one sample we have.

If someone claimed that the parameter had a value of 50, our sample result is too far away to be reasonable. If the parameter is 50 with a bounce of 3.21, how did we manage to get a parameter estimate of 23.45? That is just too much bounce for an SE of 3.21.

The third use of the SE, hypothesis testing, makes more formal use of the logic of "what is so unlikely that it is unreasonable." Without explicitly doing so, experts use the SE to get a feel for how far away the estimate is from zero (no effect). Software reinforces this by routinely adding stars for statistically significant coefficients. This will be explained in more detail when we examine hypothesis testing.

The estimate ± SE approach is a rough, rule-of-thumb guide to the bounce in the estimate. Computer software routinely displays regression coefficients along with their associated SEs. This first, most basic use of the SE gives us a feel for the variability involved and the precision in the estimate.

## 23.2 Confidence Intervals

Instead of a point estimate, using the SE to construct an *interval estimate*, a range of values, is the second way to use an SE.

The actual computation of the interval is trivial—simply subtract and add the SE of $b_1$ from the point estimate for a 68% confidence interval or first multiply the SE of $b_1$ by 2 (1.96 is a closer approximation) to get a 95% confidence interval.

Actually understanding what a confidence interval means is much more difficult than constructing it. We use our usual strategy, a concrete example, to explain confidence intervals.

*STEP* Begin by setting up a worksheet as in Figure 23.1. Use the labels to name the cells in range A1:A3. Make X from 1 to 100 by 1. Use LINEST to regress Y on X for just the first 5 observations. Later, we will use all 100 observations to understand the effect of increasing the sample size. Save the file as a macro-enabled workbook with the name, *UtilizingSE.xlsm*.

| | A | B | C | D | E | F | G |
|---|---|---|---|---|---|---|---|
| 1 | 10 | beta0 | | | | | |
| 2 | 20 | beta1 | | | | | |
| 3 | 40 | sigma | | | | | |
| 4 | | | | | | | |
| 5 | X | epsilon | Y | | estimate of beta1 with n=5 | | |
| 6 | 1 | -44.043 | -14.043 | | 29.77699 | -30.7946 | |
| 7 | 2 | -7.5115 | 42.4885 | | 11.2856 | 37.43009 | |
| 8 | 3 | -29.6767 | 40.32328 | | 0.698845 | 35.68819 | |
| 9 | 4 | 45.6543 | 135.6543 | | 6.961658 | 3 | |
| 10 | 5 | -21.741 | 88.25902 | | 8866.692 | 3820.94 | |
| 11 | 6 | 6.849113 | 136.8491 | | | | |

Figure 23.1: Implementing the DGP.

We will be using an Excel Stock chart type with High-Low-Close so we arrange the interval in cells in that order.

STEP In cell H5 enter the label "68% CI Hi" and in cell H6, enter the formula: *=E6+E7.* In cell I5 enter the label "68% CI Lo" and in cell I6, enter the formula: *=E6-E7.* In cell J5 enter the label "b1" and in cell J6, enter the formula: *=E6.* Press F9 a few times.

Cells H6 and I6 are the endpoints of the interval. They bounce when you press F9, which means the interval is a random variable.

Now, the crucial question concerns the performance of the interval: Does it cover? This means, does the interval's range contain the parameter value, $\beta_1$?

Pay close attention to the wording. It is the interval that bounces so it is the interval that does or does not cover. The parameter is a constant, fixed value.

STEP In cell K5 enter the label "Does it cover?" and in cell K6, enter the formula: *=IF(I6<beta1,IF(H6>beta1,1,0),0).* Press F9 a few times.

The formula checks if I6, the left (or min) endpoint of the interval is smaller than $\beta_1$ and, if it is, it checks if H6, the right (or max) endpoint is greater than $\beta_1$. If both of these conditions are met, the interval covers so a 1 is displayed. If either is not met, the interval does not contain the parameter so a 0 results.

A 68% confidence interval should cover 68% of the time. Does it? We can run a simulation to check, but before we do, we can create a visual display that conveys what is going on here.

To do so, we need to write some Visual Basic code. The basic idea is to have Excel resample for us and keep track of the intervals generated by each sample. We proceed slowly.

STEP Access the Visual Basic Editor by pressing alt-F11 and click *Insert* in the top menu and select *Module.*

Visual Basic displays what looks like a blank piece of paper. This is where we write code for our macros.

STEP Type the text *Sub MyFirstMacro* and hit *enter.*

The interface is sophisticated. It added "()" and "End Sub" for you. It also uses color to help you read the code.

*STEP* In the second line (where the cursor should be blinking), type an apostrophe (single quote) followed by the text *This is my first macro* and press enter.

You entered a comment and it is displayed in green text.

*STEP* Type the text *Msgbox "Hello, world!"* (ignore the context-sensitive help information) and press enter. Press F5 or click the green triangle button (under the *Debug* menu item) to run the macro.

You are returned to Excel and your message is displayed above the sheet. Clicking OK returns you to the Visual Basic Editor.

This is cool, but we want to access the macro from Excel. We will put a button on the sheet and use it to run the macro.

*STEP* Return to Excel by pressing alt-F11 (it is a toggle). Click the *Developer* tab in the Ribbon (if it is not there, click *File* → *Options* → *Customize Ribbon*), then click *Insert* and select the *Button* in the top-left corner under *Form Controls* as shown in Figure 23.2.

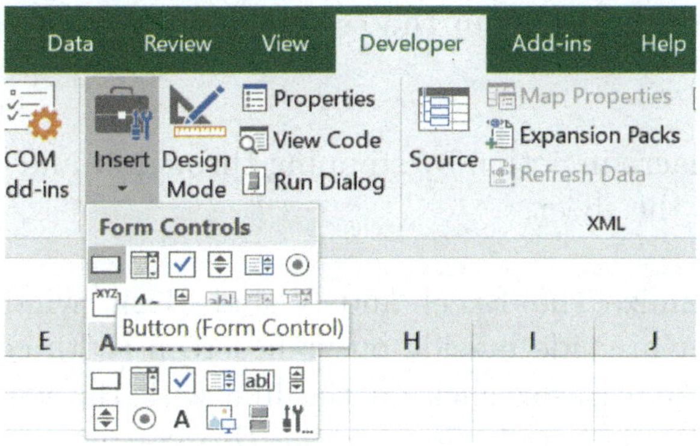


Figure 23.2: Selecting the Button (Form Control).

*STEP* When you scroll back over the sheet, the cursor is a thin cross-hair. Click and Excel displays the *Assign Macro* dialog box. Select *MyFirstMacro* and click OK.

You have attached the macro to that button. It should be selected on the screen so you can resize it if needed. But more importantly, you can give it a caption.

STEP With the button selected (right-click if needed), change the text on the button face to *Click Me.* Click on a cell in the sheet (so the button is unselected) and then (left) click the button.

Having written your first macro and run it from a button on the sheet, we proceed to more serious work.

STEP Return to the Visual Basic Editor by pressing alt-F11. Below your first macro, type the text *Sub CIConstructor* and hit *enter.* Type in these three lines of code:

```
For i = 1 To 100
Range(Cells(i, 14), Cells(i, 16)).Value = Range("H6:J6").Value
Next
```

This is a loop. On the first pass, it says to take the values in cell range H6:J6 and put them then in the three cells in row $i = 1$ and columns 14, 15, and 16 (which are N, O, and P).

When it writes the values to the sheet, Excel recalculates. On the second pass, it puts the new values in the second row because $i$=2 and so on for 100 iterations.

We can see the macro in action by stepping through it and directly observing what it writes to the sheet.

STEP Minimize the Excel and Visual Basic windows and arrange them so that they are side-by-side on your screen (with columns N, O, and P visible). In Visual Basic, click on the line with the word *Next* and then click *Debug* in the menu and select *Toggle Breakpoint.*

This sets a break in the code so the macro will stop when it reaches this line.

STEP Press F5 to run the macro.

In Excel, you can see that there are values in columns N, O, and P. They were placed there by the macro.

*STEP* Press F5 a few times and watch what happens on the spreadsheet.

Each time you press F5, the macro runs, but it stops when it reaches the breakpoint. You are stepping through the macro to see how it functions.

*STEP* Remove the breakpoint by clicking on the circle by the word Next. Press F5 and the macro runs the entire loop.

Watching how a macro operates is invaluable for debugging and understanding how Visual Basic controls what is happening on the spreadsheet.

*STEP* Return to Excel (alt-F11) and right-click your *Click Me* button. Select *Assign Macro* and assign the *CIConstructor* macro to this button. Change the button face to *Create CIs*. Click on a cell and then click the button.

Excel writes the 100 confidence intervals in High-Low-$b_1$ order in columns N, O, and P. We can use Excel's Stock chart type to display the data.

*STEP* Select columns N, O, and P, click *Insert* and click the bottom-right corner of the *Charts* group to display all of the charts. Select the *Stock* chart and *High-Low-Close* type and click *OK*.

Excel puts a chart on the sheet with 100 vertical lines with a tick mark in the middle. The length is the confidence interval and the tick mark is the point estimate, $b_1$.

*STEP* Add a horizontal red line at 20, the value of $\beta_1$. Make the max of the $y$ axis 80 and min $-40$.

Your chart should look like Figure 23.3. A few of the vertical lines are far from the horizontal red line. These are confidence intervals that did not cover.
But many of the intervals contain or overlap the horizontal red line. These intervals covered the parameter, $\beta_1$. This is good.

*STEP* Press the *Create CIs* button a few times.

The chart comes to life and the intervals bounce around every time you click the button. The horizontal red line stays fixed.

STEP Look at the chart as you click. The intervals slide up and down because each new sample has a new $b_1$ so the center of the interval shifts.

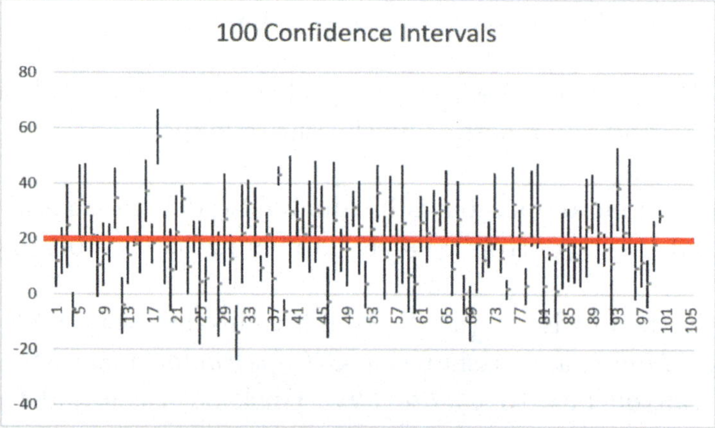


Figure 23.3: Stock Chart with High-Low-Close.

Less obvious, but also true, is that the lengths of intervals vary because the SE of $b_1$ is changing with each new sample. To be clear, the exact SE of $b_1$ is constant, but the estimated SE of $b_1$, the one we are using, is bouncing.

But is it really true that 68% of the intervals constructed by adding and subtracting the SE cover the true parameter value?

STEP Run a simulation with 10,000 repetitions of cell K6. In the Notes section, describe your results and findings.

It is immediately apparent, by looking at the average of the simulation, that these intervals are not working as advertised. They cover only about 60% of the time, not 68%. What is going on here?

It is the estimated SE of $b_1$ that is at fault. What if we use the exact SE?

STEP Enter the label "Exact SE" in cell E12 and make the text red to indicate that this is not known in a real-world setting. Enter the formula for the exact SE of $b_1$ in the cell below it: =sigma/(SQRT(5)*STDEV.P(A6:A10)).

With the exact SE of $b_1$ available, we can modify the formula for our confidence interval so it is built on the exact SE of $b_1$ instead of the estimated SE.

*STEP* Modify cells H6 and I6 so they add and subtract the exact SE in cell E13. Run a simulation with 10,000 repetitions of cell K6. In the Notes section, describe your results and findings.

Now the confidence interval is working perfectly. About 68% of the intervals covered the true parameter value. This demonstrates that it is the estimated SE that is making the interval cover too infrequently.

Unfortunately, we cannot use the exact SE in a real-world setting because we would not know sigma. The best we can do is estimate $\sigma$ with the RMSE to get an estimated SE.

Fortunately, the RMSE is a biased, but consistent estimator of $\sigma$ and, therefore, the estimated SE is a biased, but consistent estimator of the exact SE. With $n = 5$, we do not have a large enough sample size for the sampling distribution of the estimated SE to converge to the normal curve. What happens if $n$ rises, say to 100 observations?

*STEP* Copy cell range E5:K10, select cell E16, and paste. Edit the label in cell E16 to display "n=100" and edit LINEST in cell E17 so it uses all 100 observations. Press F9 a few times and compare cells E6 and E17.

With $n = 100$, there is much less bounce in $b_1$. This, of course, is really good news.

Another way of seeing the improvement in our estimate of $\beta_1$ is by watching the 68% confidence intervals.

*STEP* Modify the VBA code to capture 100 CIs based on $n = 100$. You have to change the line Range("H6:J6").Value to Range("H17:J17").Value. Return to Excel and click the button.

The confidence intervals are clumped more tightly around the parameter value's horizontal line. They are also much shorter with $n = 100$ since the estimated SE is much smaller. Are 100 observations enough to make the confidence intervals behave as expected?

*STEP* Run a sim with 10,000 repetitions that tracks cells K6 and K17.

Yes, with 100 observations, a confidence interval constructed by adding and subtracting 1 SE of $b_1$ from the point estimate of $b_1$ covers the true parameter value about 68% of the time. The normal approximation is working well here.

What if we had a small sample so we could not take advantage of the consistency of the estimated SE? In that case, we use the $t$ distribution to construct the confidence interval.

Developed by Gosset while he was working at the Guinness Brewing Company, Student's $t$ distribution is the correct distribution when errors are normally distributed and $\sigma$ is unknown.

Figure 23.4: William Sealy Gosset (1876 - 1937) generated by ChatGPT.

Guinness would not allow its scientists to use their own names so Gosset published his results under the pen name *Student*. We can use Excel's T.INV function to show how the $t$ distribution works.

*STEP* In cell K1 enter 68% and in cell L1 enter the label "Desired Confidence Level." In cell K2 enter 3 and cell L2 enter the label "degrees of freedom" (this is $n - k$ so 5 observations minus 2 coefficients is 3 degrees of freedom). In cell K3 enter the formula =NORM.S.INV(0.5+K1/2) and in cell L3 enter the label "normal" Finally, in cell K4 enter the formula =T.INV(0.5+K1/2,K2) and in cell L4 enter the label "t."

Excel displays a number very close to 1 in cell K3, but K4 is a number about 20% bigger. This bigger number is what we have to multiply the estimated SE by to make the confidence interval work correctly.

*STEP* Change the Hi and Lo values of the CI in cells H6 and I6 to =E6+K4*E7 and =E6-K4*E7. Run a sim with 10,000 repetitions that tracks cell K6.

You fixed the problem with the confidence intervals! They now cover 68% like they are supposed to. This is because you did not use the normal curve, but instead used the $t$ distribution to construct the interval.

Gosset's work only really matters if the sample size is small, by convention, defined as less than 30 observations. With more observations than that, Student's $t$ and the normal curve are essentially the same.

*STEP* Change cell K2 to 100. Compare cells K3 and K4. Increase cell K2 to 1000. Compare cells K3 and K4.

They were quite close at 100 and almost exactly the same at 1000. This means that the normal curve and the $t$ distribution will produce almost the same CI when the sample has many observations.

To be clear, the $t$ distribution is always the exactly correct distribution to use with normal errors and unknown $\sigma$. Regression software will show confidence intervals (and tests of significance) based on the $t$ distribution because it assumes the errors are normal and $\sigma$ is unknown. It will compute confidence intervals correctly so that if $n$ is small, the intervals are longer than what the normal curve would produce.

Software uses the $t$ distribution, no matter the sample size, because that is the correct distribution for normal errors and unknown $\sigma$. As $n$ increases, the $t$ distribution rapidly approaches the normal distribution. So using the $t$ distribution only matters for small sample data sets, say less than 30 observations.

It is easy to get lost in the complications brought on by small-sample statistical theory. In practice, regression software will correctly compute confidence intervals. The bigger issue is understanding what the range is telling us.

*STEP* In cells H17 and I17, change the formulas so the interval adds and subtracts 1.96 times the SE of $b_1$. Run a sim of cell K17.

This simulation shows clearly that about 95% of intervals constructed by adding and subtracting 2 SEs from the point estimate will cover.

A 95% confidence interval says that 95% of the intervals will contain the parameter. That is, we can be 95% sure the range includes the parameter.

The wording matters here. It is wrong to say that a 95% confidence interval means that "there is a 95% chance that the parameter is in the interval." That makes it seem like the parameter is bouncing around and 95% of the time it falls in the interval. This is false.

Constructing confidence intervals in the spreadsheet make it clear that it is the interval that bounces. The proper wording is that "there is a 95% chance that the interval covers" or "we are 95% confident that the interval covers." Remember, the randomness is in the interval, not the parameter.

By convention, 95% confidence intervals are usually reported. We can produce 99% confidence intervals by constructing an interval from $b_1 - 3SE$ to $b_1 + 3SE$.

We could also create a 100% confidence interval by having the range go from negative infinity to positive infinity. That interval will certainly cover the parameter value. Although true, that is not very helpful.

## 23.3 Takeaways

The first way to convey variability in the estimate is the easiest: simply report the estimate with its SE. The knowledgeable reader provides her own interpretation, recognizing that the estimate is a random variable and the SE indicates its spread.

There are two ways to present regression coefficients with their SEs. With only a few variables, an equation with SEs in parentheses, under the coefficients, can be used. If there are many variables, a table is a better option.

The second way of communicating variability in an estimate, a confidence interval, requires less work by the reader. Instead of a single number, a range is given, along with a level of confidence in the interval.

There are, however, dangers in this approach because confidence intervals can be misunderstood. It is untrue that a confidence interval tells you the chances that a parameter is in the interval. Thus, it is wrong to say that a 95% confidence interval means that there is a 95% chance that the parameter is in the interval.

The parameter is fixed and constant. There is no chance associated with it. Do not connect the parameter to chance or probability.

Here is another example of the wording mistake. The detective says, "Given the evidence we have, there is a 95% chance he is guilty." That is really sloppy language. He is either guilty or not. There is no chance involved. The uncertainty is in the detective's mind. Better wording is, "The evidence we have would correctly identify the guilty person 95% of the time."

It is the interval that is random. A 68% confidence interval is constructed by using the SE so that 68% of the intervals cover the parameter value. If $n$ is large and the sampling distribution follows the normal curve, $b_1 - SE$ to $b_1 + SE$ will cover about 68% of the time. About 95% of intervals constructed by $b_1 - 2SE$ to $b_1 + 2SE$ will cover.

# 24. Hypothesis Testing

The third and final way to use the SE is in a hypothesis test. As with confidence intervals, subtle wording is at play and there is deeply embedded confusion about the meaning of terms like "statistical significance."

We begin by running a hypothesis test on $b_1$ with a claim (null hypothesis) that $\beta_1 = 20$. This claim happens to be true, the parameter really is 20, but in the real world, we do not know that. We are interested in using our estimate to decide if the claim about the parameter value is true or not.

The logic is contrarian. Our default is that the claim is true, like an accused person on trial is innocent until proven guilty. Our estimate is the evidence.

## 24.1 Creating a Test Statistic

Here is an idea: if our estimate is 20, then the claim is true; if not, it is false. We can implement this procedure in Excel.

*STEP* In cell E2 enter the formula: *=E6-beta1*. Press F9 a few times.

The problem with this idea is immediately apparent: $b_1$ is never going to be exactly 20. Our estimate has bounce embedded in it so it hops around. How can we decide if our estimate is close enough that it might have reasonably been produced by a $\beta_1 = 20$ value?

A straightforward approach is to divide the difference by the SE of $b_1$. To be clear, we use the *estimated* SE of $b_1$ since that is the only one of the three SEs that we can get. This produces a difference in standardized units which enables us to judge how close or far the estimate is from the claim.

If we get an estimate of 25, the difference from 20 is 5. We have no way of knowing if this is close or far from the claim. If we know the SE is 10, we can say that 5 is 0.5 in SE, or standard deviation units, units and conclude that this is not very far away.

More formally, we create what is called a *test statistic* by computing the difference between our estimate and the claimed parameter value in SE units, that is, divided by the SE of $b_1$:

$$b_1 \text{ test statistic} = \frac{b_1 - \text{ claim}}{\text{estimated SE of } b_1}$$

*STEP* In cell E3 enter a formula for the $b_1$ test statistic for the claim that $\beta_1 = 20$. Press F9 a few times.

Not surprisingly, cell E3 is a random variable. After all, it has a random variable in the numerator, $b_1$, and a random variable in the denominator, the estimated SE of $b_1$.

*STEP* Run a simulation of cell E3 with 10,000 repetitions.

There is a lot of bounce in the $b_1$ test statistic so you may get an average that seems far from zero, even with 10,000 repetitions. Since the claim is true, however, subtracting it from the unbiased slope estimator means the expected value of our $b_1$ test statistic really is zero.

But what about its shape? We can superimpose the normal curve on the simulation histogram to see if our $b_1$ test statistic is normally distributed.

*STEP* Scroll right and, in cell AM1 enter the label "N CDF." In cell AM2, enter the formula: *=NORM.DIST(AK2,$J$5,$J$6,1)*. Fill it down.

The *cumulative distribution function*, CDF, for each bin value is in column AK. The CDF tells us the probability that a random variable is less than or equal to a given value. For a normally distributed random variable, the chances that it is less than or equal to zero, in standard units, is 50%.

To compute the number expected in each bin, we can take the difference in the CDF from one bin to the next and multiply by 10,000. The bins are repeated in column AK to draw the horizontal part at the top of each bin in the chart so we have to account for this in our formula.

*STEP* In cell AN1, enter the label "N prob" and in cell AN2, enter the formula: *=(AM4-AM2)*10000*. Fill it down to the second cell from the bottom, AN81. You should see pairs of repeated values.

Next, we need to modify the $x$ data for the SERIES formula.

STEP In cell AO2 enter the formula: *=AK2+0.5*. Fill it down. Return to the chart and superimpose the normal curve by copying the existing SERIES formula and changing AK to AO and AL to AN.

Figure 24.4 shows the results of a simulation of the $b_1$ test statistic with a superimposed normal curve. The fit of the curve is poor because the $b_1$ test statistic does not follow the normal curve. While $b_1$ is unbiased and normally distributed, dividing it by the estimated SE complicates matters.

| Summary Statistics | | Notes |
|---|---|---|
| Average | -0.029 | b1 test stat does not follow the normal curve; in fact, its distribution is Student's *t* |
| SD | 1.6459 | |
| Max | 21.065 | |
| Min | -17.588 | |

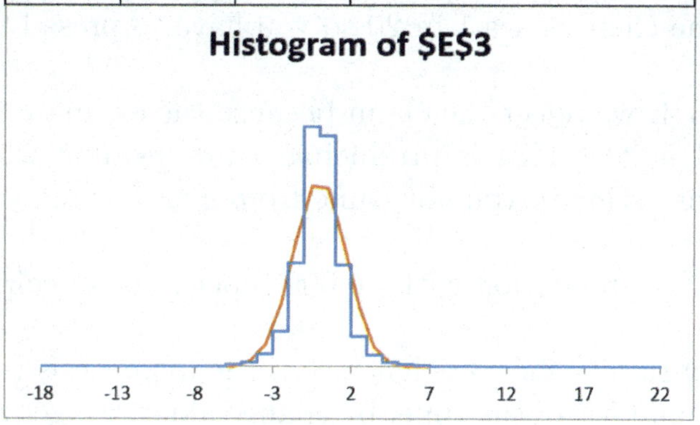


Figure 24.1: Simulating the $b_1$ test statistic.

## 24.2 Gosset's Contribution

What Gosset figured out (without simulation) is that the normally distributed numerator ($b_1$ − *claim*) divided by the estimated SE (which is another random variable) is not normally distributed and this is especially noticeable when the sample size is small. If the errors are normally distributed, Gosset discovered the distribution of the test statistic and named it Student's $t$ distribution. This is why this hypothesis test is called a $t$ test.

Gosset also understood that the test statistic does go to the normal curve as $n$ rises. The $t$ distribution only matters if the sample size is small. As $n$ rises, the $t$ distribution rapidly converges to the normal curve.

Once we have our test statistic, we can compute its *probability value*, which we abbreviate as $p$-value. We cannot compute the chances of getting a single, specific estimate since that would be effectively zero, so we compute the chances of getting a range from our estimate and higher.

STEP Return to the *DGP* sheet to compute the $p$-value. In cell E4 enter the formula: *=T.DIST.2T(ABS(E3),F9)*. Format as % with one decimal point. Press F9 a few times.

Cell E4 is showing the probability of getting our estimate or one more extreme, assuming the claim is true, from either side of the sampling distribution (the 2T in Excel's T.DIST function is for two tails). We reject the claim if the $p$-value is low, less than 5% by convention.

STEP In cell F4 enter the formula: *=IF(E4<0.05,1,0)*. Press F9 until you get a 1. The chances are 1 in 20 so you have to press F9 quite a bit.

When cell F4 = 1, we reject the claim because the estimate is far away from 20 in standard units. This is producing a low $p$-value which means it is unlikely that the estimate you got came from a true claim.

STEP Run a simulation with 10,000 repetitions of cell F4.

Your simulation results show evidence that the hypothesis test is working as advertised. Around 5% of the time, by chance alone, we get an estimate that seems too far away to be reasonable and we call this *statistically significant*.

The term *highly statistically significant* is used for $p$-values less than 1%. You might think this is a good idea since we are less likely to reject a true claim (like $\beta_1 = 20$). However, if the claim is false, then a 1% criterion means we fail to reject more false claims.

It is absolutely crucial that you understand that statistically significant does not mean *practically important*. Because significant is synonymous with important in everyday language, it is easy to think that statistical significance means the result really matters or is especially meaningful. It does not.

Instead, statistical significance tells us that the estimate is far enough away from a claimed parameter value that we do not believe the claim because chance alone is unlikely to have caused the difference. That is all it says.

It would be clearer if we always used the phrase, "statistically significantly different from the claim," but no one does this.

It is common to interpret statistical significance as a substantial result—so much so that some experts want to ban the language altogether (Wasserstein & Lazar, 2016).

We have been working with what we know is a true claim, $\beta_1$ really does equal 20, but that is not how hypothesis testing works in the real world. In practice, we do not know the parameter value and the usual claim is that the parameter value is zero.

In a regression setting, the claim that a slope parameter value ($\beta_k$) is zero means that the independent variable ($x_k$) has no effect on the dependent variable ($y$). This is a *straw man argument*—we want to reject the claim and declare that we have a statistically significant result.

*STEP* Change the formula in cell E3 to test the claim that $\beta_1 = 0$. In other words, make the formula =(E6-0)/E7.

Most of the time, the $p$-value in cell E4 is greater than 5%. This means that we would not have enough evidence to reject this claim with the sample evidence we have, even though it is false. The small sample size results in a large SE, which means that the estimated coefficient is imprecisely estimated. Not surprisingly, we cannot reject the false null.

What happens if we test this claim with 100 observations?

*STEP* Copy cells E3:F4 and paste in cell E14.

The $t$ stat is huge and the $p$-value almost zero. This means that the chances of a sample estimate like the one you got being produced by an equation where $\beta_1 = 0$ is absurdly small. Therefore, we would reject this claim.

Be sure to notice, again, that the SE for $n = 100$ is way smaller than $n = 5$. Our ability to so much more precisely estimate $\beta_1$ is the reason why we can reject the claim that $\beta_1$ is zero.

Since we understand confidence intervals, we can use them to give insight into what hypothesis testing is all about. To run a test of significance, simply create a 95% confidence interval by computing the range from

$b_k - 2 * SE(b_k)$ to $b_k + 2 * SE(b_k)$ and ask if it covers the claimed $\beta_k$. If it does not, the result is statistically significant at the 95% level.

If we create a 99% confidence interval, using $\pm$ 3SEs, and determine if it covers the claimed value, we can declare if our estimate is highly statistically significant or not.

The effort required to master hypothesis testing has a low rate of return when working with huge data sets like the CPS because the sample size is so large. This makes the SE really small and generates teeny $p$-values for nulls of no effect. Thus, we are more interested in estimating the magnitude of the effect rather than making an either-or pronouncement about statistical significance.

You know that a teeny $p$-value does not mean we have an important or meaningful result, but you should also be aware that statistical significance tells us nothing about omitted variable bias. And even more distressing, increasing the sample size does nothing to fix omitted variable bias.

## 24.3 Takeaways

There are three ways to use an SE:

1. As a give or take, plus-minus number on the estimated parameter.
2. To create a confidence interval (CI) estimate for the parameter.
3. To conduct a hypothesis test of a claimed value of the parameter.

This chapter focused on the third way of using the SE, hypothesis testing. This is a complicated and often frustrating area of statistics. The language is contorted and it is easy to misinterpret the result.

The basic idea is to determine if the estimate is far enough away from a claimed parameter value to warrant rejection of the claim.

We introduced Monte Carlo simulation with a free throw shooting example. A 90% free throw shooter makes only 75 out of 100 free throws. Do we believe that they are truly a 90% shooter and just had a lot of bad luck in 100 attempts?

Simulation showed that making only 75 or fewer out of 100 is incredibly rare so we would reject the claim. Hypothesis testing reaches the same conclusion, but the method is completely different.

The SE is 3 so 75 minus 90 divided by 3 means that we are 5 SEs away from what we expected. This is incredibly unlikely. The $p$-value would be nearly zero which means that it is really rare to get a result of 75 or fewer if the sampling distribution is centered on 90%. Thus, we reject claim that the sampling distribution is centered on 90%.

Many experts routinely divide the slope coefficient by its SE and use 2 as a guide. This is basically a quick hypothesis test of a claim that the slope is zero. A result bigger than 2 means the estimate is statistically significant.

There are many ways to misinterpret hypothesis tests, but two big mistakes to avoid are:

1. Believing that statistical significance measures the size of an effect or indicates the importance of a result.

2. Interpreting a $p$-value as the chances that a claim is true.

The logic of hypothesis testing extends beyond the simple, one-parameter hypothesis test. The idea of using a single result from a random process as evidence has many applications. The problem is figuring out how to determine if the single result is "far enough away" to reject the claim.

## References

Wasserstein, R. & Lazar, N. (2016) "The ASA's Statement on $p$-Values: Context, Process, and Purpose," *The American Statistician*, 70(2), 129-133. www.tandfonline.com/doi/full/10.1080/00031305.2016.1154108

# 25. Joint Hypothesis Testing

A $b_k$ test statistic can be used to decide if our slope estimate is far enough away from a claim about $\beta_k$ to reject the claim. This is a basic, single hypothesis test, but there are more sophisticated ways to use sample evidence to judge claims about relationships.

A method worth mastering is the *joint hypothesis test.* It allows us to go beyond looking at a single variable's effect so we can evaluate more complicated claims about how variables might be interacting.

This section makes clear the common logic underlying all joint hypothesis testing, of which there are many variants. After reviewing $\chi^2$ (chi-square) and $F$ distributions, we explore three examples with hypothetical data:

1. Does experience matter in an earnings function?
2. Are SNAP benefits and cash the same?
3. The Whole Model $F$ test, a diagnostic statistic of the overall performance of the model.

We will take advantage of our ability to implement the data generating process and know parameter values to run simulations that reveal how a joint hypothesis test works. Along the way, we will learn about two new distributions with Excel functions, *CHISQ* and *F*.

## 25.1 Two New Distributions

We have drawn random numbers many times. We are quite familiar with the uniform distribution through Excel's native RAND() function and the UNIFORM(mean, SD) function in the MCSim add-in. We have also used NORMALRANDOM(mean, SD) and Excel's native NORM.INV() function repeatedly to draw normally-distributed errors. And, we have used BIVARNORMAL() (in the MCSim add-in) to draw correlated, normally distributed random numbers.

We know that Gosset figured out that a normally-distributed random variable (like an average or regression slope) divided by its estimated SE follows Student's $t$ distribution (which converges to the normal curve as $n$ increases).

But this is just the tip of the iceberg. There are many more distributions. Two that are relevant for our work with joint hypothesis testing are the $\chi^2$ and $F$ distributions. With Excel, we can take them out of the abstract, imaginary world of formal mathematics and make them clearly visible.

*STEP* In cell A1 enter the formula: *=NORMALRANDOM(0,1)*. Run a simulation with 10,000 repetitions.

You get an approximation to the familiar normal curve. What happens if we add two N(0,1) random variables?

*STEP* In cell A2 enter the formula: *=NORMALRANDOM(0,1)* and in cell A3, add the two cells. Run a simulation with 10,000 repetitions of cell A3.

Summing two normally-distributed random variables increases the spread, but the sum remains normally distributed.

What happens if we square a N(0,1) random variable? That is, we multiply them instead of adding them?

*STEP* In cell B1 enter the formula: *=NORMALRANDOM(0,1)^2*. Run a simulation with 10,000 repetitions.

You are looking at a $\chi^2$ distribution with 1 degree of freedom, df. The uniform and normal distributions depend on the mean and SD, but the $\chi^2$ distribution depends just on a df value. It gives the center of the distribution and controls its shape.

The $\chi^2$ distribution cannot be negative (squaring guarantees that) and it has many values near zero (the tall rectangle). This is because there are many numbers near zero in N(0,1) so when they are squared, this produces even more numbers near zero.

The $\chi^2$ distribution also has a long tail because draws from the tail of N(0,1) (greater than 2 in absolute value), get squared and, thus, magnified.

We can produce random numbers from such a distribution by embedding RAND() in Excel's CHISQ.INV function.

*STEP* In cell C1 enter the formula: *=CHISQ.INV(RAND(),1)*. Run a simulation with 10,000 repetitions of cells B1 and C1.

Your sim shows that the two cells are the same (you can increase the number of repetitions if you doubt it). The strategy of embedding RAND() in an INV (inverse) distribution function works for any distribution.

STEP Modify your $\chi^2$ distribution formula in cell C1 so that it has 10 df: *=CHISQ.INV(RAND(),10)*. Run a simulation with 10,000 repetitions.

Now it looks more bell-shaped while centered on 10. This shows that the shape and location of distributions depend on parameters.

Can you guess what we would have to do with NORMALRANDOM to get a $\chi^2$ with 10 degrees of freedom? See the Appendix to check your answer.

Our real interest, however, is not in $\chi^2$, but the $F$ distribution, which is named after one of the most important statisticians in history, Ronald Fisher. The $F$ distribution is produced when you take the *ratio* of two $\chi^2$ distributed random variables.

STEP In cell B2 enter the formula: *=NORMALRANDOM(0,1)^2* and in cell B3, enter the formula: *=B1/B2*. Run a simulation with 10,000 repetitions.

The result looks like a single tall rectangle. Actually, it does have a long tail, but there are very few observations beyond the mass near zero.

Notice that we are dividing a $\chi^2$ with 1 df by another $\chi^2$ with 1 df to get this F distribution. Both $\chi^2$ squares have a tall rectangle and a long tail and then we divide one by the other to generate the F distribution.

Like every distribution, the $F$ depends on its parameters, of which there are two: df in the numerator and df in the denominator. Figure 25.1 shows the shape of the $F$ for a few examples, including our simulation (d1=1, d2=1).

We can easily reproduce the d1=100, d2=100 light gray $F$ distribution in Figure 25.1.

STEP In cells G1 and G2 enter the same formula, *=CHISQ.INV(RAND(),100)*. Now take the ratio of the two chi-squared random variables in cell G3 by entering the formula *=G1/G2*. In cell H3, enter the formula, *=F.INV(RAND(),100,100)*. Finally, sim cells G3 and H3.

Your sim results show they are the same (the averages and SDs are quite close and the histograms are roughly superimposed on top of each other). The distributions produced by simulating cells G3 and H3 are also quite close to the d1=100, d2=100 light gray curve in Figure 25.1.

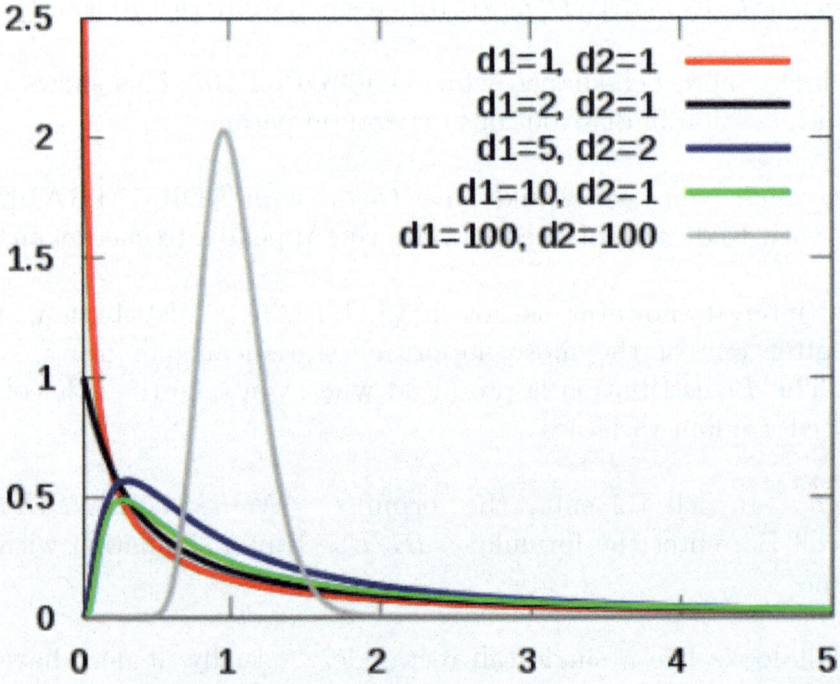


Figure 25.1: The $F$ Distribution with Different Degrees of Freedom. Source: en.wikipedia.org/wiki/F-distribution

## 25.2 Does Experience Matter?

We worked on learning two new distributions because our joint hypothesis test is based on the $F$ distribution, which comes from the $\chi^2$ distribution. Our joint hypothesis test compares the SSR of two regressions and we decide if they are far enough apart to declare them different.

SSRs are $\chi^2$ distributed since we square normally-distributed residuals (we assume $\epsilon_i \sim N(0, \sigma)$) and add them up. The comparison of the SSRs is based on a percentage change with SSR in the numerator being divided by SSR in the denominator. This gives an $F$ distribution.

To perform a joint hypothesis test, we need to compare *unrestricted* and *restricted* regressions. The unrestricted regression is the full model. It always has a lower SSR. The restricted regression has a higher SSR because it uses fewer independent variables (this is how the restriction is incorporated) and the question then becomes: is the loss of fit (measured by bigger SSR) large enough to decide that the restriction matters?

To show how it works, we create our first example. Recall that an earnings function regression explains some measure of earnings, like hourly wage or annual salary, as a function of several independent variables.

The effect of education (the number of years of schooling) on earnings is often the focus. A natural log transformation of the earnings data is typically used to capture the fact that education produces a multiplicative increase in earnings. We ignore this to make it easier to interpret the coefficients.

Instead of education, we are interested in the effect of experience on earnings. Recall that experience has a positive effect on earnings at first, but more experience drags down earnings for older workers. We capture this in an earnings function by using experience and experience squared.

Thus, our model is:

$$Earnings_i = \beta_0 + \beta_1 Education + \beta_2 Experience + \beta_3 Experience^2 + \epsilon_i$$

The $\epsilon_i$ are normally distributed with mean zero and SD of sigma. They are iid and we will make the $x$ variables be fixed in repeated sampling to exactly meet all of the requirements for a well-behaved DGP.

*STEP* Create a sheet based on Figure 25.2. Name the cells in range A1:A5 with their labels. For education, enter the formula to draw years of schooling from 12 to 16. Epsilon is N(0,sigma) and earnings in cell E9 has the formula: *=intercept+slope_education*A9+slope_experience*B9+slope_experience2* C9+D9*. Fill it down. There are 41 observations.

*STEP* Copy the sheet and deaden the Xs (copy and paste values, switching calculation to manual if needed). Use LINEST in cells G9:J13 to run a regression of earnings on education, experience, and experience$^2$. Press F9 a few times.

This is the unrestricted regression and it is producing reasonable results. The estimates are bouncing around their parameter values.

| | A | B | C | D | E |
|---|---|---|---|---|---|
| 1 | 50000 | intercept | | | |
| 2 | 2000 | slope education | | | |
| 3 | 2500 | slope experience | | | |
| 4 | -50 | slope experience$^2$ | | | |
| 5 | 10000 | sigma | | | |
| 6 | | | | | |
| 7 | =ROUND(RAND()*5+11.5,0) | | | | |
| 8 | education | experience | experience$^2$ | epsilon | earnings |
| 9 | 16 | 0 | 0 | 14529.06 | 96529.06 |
| 10 | 13 | 1 | 1 | -2322.08 | 76127.92 |
| 11 | 14 | 2 | 4 | -4283.75 | 78516.25 |

Figure 25.2: Experience Hypothesis Test Setup.

But here is the key question: if we wanted to test the claim that experience does not matter, how would we do it?

We cannot just create a $b_2$ test statistic, dividing the estimated slope on experience by its SE, because experience is being modeled as experience and experience$^2$. If experience does not matter, then the claim is that *both* $\beta_2$ and $\beta_3$ are zero. How do we handle that joint claim?

We run a joint hypothesis test. We apply the claim and then see if the loss of fit convinces us that the model is different without experience.

*STEP* Use LINEST in cells L9:M13 to run a regression of earnings on education alone. Press F9 a few times.

This is the restricted regression. We imposed the claim that experience does not matter by dropping both experience variables. The SSR of the restricted regression is larger, reflecting a loss of fit, but the question is, by how much?

If we get a much poorer fit, then that means experience matters and the claim is false. On the other hand, if SSR in the restricted regression is only a little bigger than SSR in the unrestricted regression, then chance is a reasonable explanation and the claim would not be rejected.

One way to measure the diminished explanatory power of the restricted model is by computing the percentage increase in the SSR.

*STEP* In cell I15 enter, *=(M13-H13)/H13*. Press F9 a few times.

The SSR is much larger in percentage terms. This means the restricted model is performing much worse. Experience, therefore, matters. This is the logic of the joint hypothesis test.

To get the test exactly right, however, Fisher figured out that we have to make a couple of technical adjustments and compute the $F$ statistic like this:

$$\frac{\frac{\text{Restricted SSR}-\text{Unrestricted SSR}}{\text{Number of Constraints}}}{\frac{\text{Unrestricted SSR}}{\text{Number of Observations}-\text{Number of Parameters Estimated}}}$$

Notice how it is essentially a percentage change computation (the numerators) with the degrees of freedom parameters as adjustments (the denominators).

*STEP* In cell M15 enter the label "numerator" and in cell L15, enter the formula: *=(M13-H13)/2*. In cell M16 enter the label "denominator" and in cell L16, enter the formula: *=H13/(41-4)*. In cell M17 enter the label "F statistic" and in cell L17, enter the formula: *=L15/L16*.

We are almost finished with our joint hypothesis test. The last step is to compute the $p$-value. Unlike the normal distribution, for which we know that more than 2 SEs away is rare, we have no rule of thumb for what is a small or large $F$ statistic.

We can compute the $p$-value using Excel's F.DIST.RT (for right tail) function. We have to provide the $F$ statistic, numerator df, and denominator df.

*STEP* In cell M18 enter the label "p-value" and in cell L18, enter the formula: *=F.DIST.RT(L17,2,H12)*. Press F9 a few times.

The $p$-value is always tiny. This is good news. It means we reject the claim that experience does not matter because if the claim were true, the chances of getting an $F$ statistic like the one we got or one even bigger, is minuscule.

What if the claim were true? How would the model and the joint hypothesis test behave in that case?

*STEP* Change cells A3 and A4 to zero. Press F9 a few times.

The change is dramatic. Now the percentage loss of fit in cell I15 is usually pretty low and the *p*-value is usually pretty high. In fact, at the 5% level for statistical significance, we will reject this true claim only 1 in 20 times.

## 25.3 SNAP Benefits versus Cash

In the United States, the Supplemental Nutritional Assistance Program (SNAP, www.fns.usda.gov/snap/supplemental-nutrition-assistance-program) provides support for needy families to buy food.

Unlike cash, SNAP benefits can only be used to buy food. The question then becomes, do people treat SNAP benefits the same as cash?

Our unrestricted model is:

$$Food_i = \beta_0 + \beta_1 SNAP + \beta_2 Cash + \epsilon_i$$

The epsilons are normally distributed with mean zero and SD of sigma. They are iid and we will make the exes be fixed in repeated sampling to exactly meet all of the requirements for a well-behaved DGP.

*STEP* Create a sheet based on Figure 25.3. Name the cells in range A1:A4 with their labels. There are 20 observations. For SNAP, make the first ten have a value of 50 and the second ten have a value of 0. For Cash, create a series from 0 to 190 by 10. Epsilon is N(0,sigma) and Food in cell E9 has the formula: *=intercept+slope_SNAP*A7+slope_Cash*B7+C7*. Fill it down. Format cells in the range A7:D26 as $.

The SNAP program does not actually work this way. We have everyone below a Cash income level of $100 getting $50 in SNAP benefits. There are eligibility criteria, but they include age, income, and number of children in the family. The implementation does capture, however, that SNAP works by providing support to needy families.

| | A | B | C | D |
|---|---|---|---|---|
| 1 | | 50 | intercept | |
| 2 | | 1 | slope SNAP | |
| 3 | | 1 | slope Cash | |
| 4 | | 30 | sigma | |
| 5 | | | | |
| 6 | SNAP | Cash | epsilon | Food |
| 7 | $ 50.00 | $ - | $ 9.28 | $ 109.28 |
| 8 | $ 50.00 | $ 10.00 | $ (15.16) | $ 94.84 |
| 9 | $ 50.00 | $ 20.00 | $ 39.13 | $ 159.13 |

Figure 25.3: SNAP versus Cash Hypothesis Test Setup.

*STEP* Use LINEST in cells F7:H11 to run a regression of Food on SNAP and Cash. Press F9 a few times.

This is the unrestricted regression and it is producing reasonable results. The estimates are bouncing around their parameter values.

But here is the key question: if we wanted to test the claim that people treat SNAP and Cash as the same, how can we do it?

The trick is to notice the implication of the claim. If the claim is true, then the slope coefficients on SNAP and Cash must be the same: slope SNAP = slope Cash.

The slopes are interpreted as how much of an additional dollar is spent on food. If they are both 1, then all of an additional dollar of SNAP or Cash is spent on food.

A little algebra gets us to the restricted regression. Since the slopes are the same, we can call them both $\beta_1$ and then combine like terms:

$$Food_i = \beta_0 + \beta_1 SNAP + \beta_1 Cash + \epsilon_i$$

$$Food_i = \beta_0 + \beta_1 (SNAP + Cash) + \epsilon_i$$

Our restricted regression imposes the restriction that the slopes are the same. To run this regression, we create a new variable that is the sum of SNAP and Cash.

STEP In cell J6 enter the label "SNAP+Cash" and in cell J7, enter the formula: *=A7+B7*. Fill it down. Select range L7:M11, enter the formula, *=LINEST(D7:D26,J7:J26,1,1)*, and hit ctrl-shift-enter.

We will run a joint hypothesis test to see if the loss of fit convinces us that the model is different with the common slope. We use the same $F$ statistic as before, computing an adjusted percentage change in loss of fit.

STEP In cell M13 enter the label "numerator" and in cell L13, enter the formula: *(M11-G11)/1*. We have only one constraint. In cell M14 enter the label "denominator" and in cell L14, enter the formula: *=G11/(20-3)*. In cell M15 enter the label "F statistic" and in cell L15, enter the formula: *=L13/L14*. In cell M16 enter the label "p-value" and in cell L16, enter the formula: *=F.DIST.RT(L15,1,G10)*. Press F9 a few times.

In our first example, we found a tiny $p$-value so we rejected the false claim that experience did not matter. This time, the $p$-value is usually quite high. We often do not reject the claim. This is good news since the claim is true—the slopes really are the same.

How often will we make a mistake and reject the true claim?

STEP In cell M17 enter the label "reject claim" and in cell L17, enter the formula: *=IF(L16<0.05,1,0)*. Sim cell L17 with 10,000 repetitions.

You should find that, as advertised, that statistical significance produces a 5% rejection rate when the claim is true.

What if the claim is false?

STEP Change cell A3 to 0.5 and run a simulation of cell L17.

We correctly reject the false null about 25% of the time.

Are SNAP benefits treated the same as cash in the real world? The answer is complicated. Whitmore (2002) runs a controlled-experiment and finds that most people's purchases are unaffected by benefits that can only be used to purchase food. Roughly a quarter, however, are distorted—they spend less on food if they get cash instead of SNAP benefits (known as food stamps when the paper was written).

## 25.4 Takeaways

A joint hypothesis test has many flavors and applications. All are based on the idea that a restriction degrades the performance of the fitted model, as measured by the SSR.

The question then becomes: is the loss of fit severe enough that we reject the restricted model?

We use the F distribution (a ratio of two $\chi^2$ random variables) because the SSR is $\chi^2$ distributed and we are working with a ratio of SSRs, properly adjusted for its degrees of freedom.

The F statistic itself is not very informative. With the normal distribution, we know that there is about a 2.5% chance of getting a value of 2 or above. There is nothing like that with the F distribution. To interpret the F statistic, we need to compute the $p$-value.

*EXCEL TIP* To compute the $p$-value for the F statistic, use this Excel formula: *=F.DIST.RT(F statistic, df numerator, df denominator)*.

Like any test of significance, it is difficult to properly interpret the $p$-value. This is known as the observed significance level and it is natural to express it as the chances that the claim is true. Unfortunately, this is poorly worded and incorrect.

To avoid this misinterpretation, remember that the $p$-value is based on the claim being true. The correct interpretation is that the $p$-value tells us the chances we would have gotten a result like we did (or one more extreme) assuming that the claim is true.

When the $p$-value is small, it is evidence against the claim because it means the sample we got was unlikely.

Remember also that this has nothing to do with practical importance. All you are testing is whether or not chance alone could have produced the observed result.

Tests of significance are really easily misunderstood.

Fisher was an avid pipe smoker and he refused to believe that smoking caused cancer as the evidence mounted in the 1950s. Maybe that is why ChatGPT depicted him with a pipe.

Figure 25.4: Ronald Fisher (1890-1962) generated by ChatGPT.

## References

Whitmore, D. (2002) "What Are Food Stamps Worth?" Working Paper #468, Princeton University. Industrial Relations Section, arks.princeton.edu/ark:/88435/dsp01z603qx42c.

Hoynes, H. & Whitmore Schazenbach, D. (2009) "Consumption Responses to In-Kind Transfers: Evidence from the Introduction of the Food Stamp Program" *American Economic Journal: Applied Economics*, (1, 4), 109-139, https://www.jstor.org/stable/25760184.

## Appendix

A $\chi^2$ with 10 df is simply the sum of 10 squared normal draws. You can check this by filling down your NORMALRANDOM(0,1)ˆ2 formula in cell B1 to B10, then summing the 10 numbers in cell B11. Check to make sure that cell C1 is CHISQ.INV with 10 df and run a sim of cells B11 and C1. Your sim results will show that they are the same.

# 26. The Whole Model F Test

The last example of a joint hypothesis test is the Whole Model F test. This common regression diagnostic is reported by LINEST. We can see how it works in both the earnings function and SNAP benefits examples.

## 26.1 Whole Model F on Earnings Function

*STEP* Return to the earnings function regression sheet and highlight cell G12 with a yellow background.

LINEST is reporting the Whole Model F statistic. To give it meaning, we must compute its *p*-value.

*STEP* In cell F12 enter the formula: *=F.DIST.RT(G12,3,H12)*. Press F9 a few times.

Excel displays the chances of getting an F statistic like the one in cell G12 or bigger. The *p*-value is typically extremely small as you press F9 so we reject the claim. This is good news since we know the claim is false.

But what exactly is the claim in the Whole Model F test? The claim is that *none* of the independent variables matter. This means that each of the slopes in the DGP are zero:

$$\beta_1 = \beta_2 = ... = \beta_k = 0$$

The claim says that the independent variables in the model have no explanatory power. We could do just as well with no $X$ variables at all.

The restricted regression in the Whole Model F test is one with no independent variables and just an intercept term, like this:

$$y_i = \beta_0$$

We employ a nifty trick: we interpret the intercept term as being multiplied by 1. This treats the intercept as a coefficient on an $x$ variable.

*STEP* In cell O8 enter the label "intercept x" and enter the number 1 in cell O9. Fill it down to cell O49.

We use this column of 1s as a single regressor with no intercept term.

STEP Select cell range Q9:R13 and enter the LINEST formula with a 0 in the third argument to suppress the intercept: *=LINEST(E9:E49,O9:O49,0,1)*. Press ctrl-shift-enter.

This restricted regression's SSR is the claimed true model. The question is whether the unrestricted regression has enough improvement, as measured by its lower SSR, to make us believe the claim is false.

We can better understand what the restricted regression is doing with a few simple computations.

STEP In cell T8 enter the label "y - avg y" and in cell T9, enter the formula: *=E9-AVERAGE($E$9:$E$49)*.

This is the deviation of actual $y$ from its mean. We can also think of it as a residual if we predicted $y$ with just its average. In other words, we would use no other explanatory variables, just like the claim said, to predict $y$.

STEP In cell U8 enter the label "squared residual" and in cell U9, enter the formula: *=T9^2*. In cell V8 enter the label "SSR" and in cell V9, enter the formula: *=SUM(U9:U49)*. Press F9 a few times.

Cell V9 is identical to cell R13, the SSR of the restricted regression with just an intercept term. This helps us understand that the restricted regression is predicting $y$ with no information other than $y$ itself.

We can now compute the F statistic with the SSRs of the unrestricted and restricted regressions. We use 3 df in the numerator (for the three slopes that are claimed to be equal to zero) and 37 df in the denominator (41 observations minus 4 parameters estimated in the unrestricted regression).

STEP In cell R15 enter the label "numerator" and in cell Q15, enter the formula: *=(R13-H13)/3*. In cell R16 enter the label "denominator" and in cell Q16, enter the formula: *=H13/H12*. In cell R17 enter the label "F statistic" and in cell Q17, enter the formula: *=Q15/Q16*. Hit F9 a few times.

We have exactly replicated the LINEST F statistic in cell G12. Our work shows that LINEST is displaying a Whole Model F statistic and demonstrates how this joint hypothesis test works.

Now that we know what F statistic LINEST is displaying, what can we do with it? By computing its $p$-value (which we have already done in cell F12), we can decide whether or not to reject the claim that none of the explanatory variables matter.

The $p$-value is giving us the area in the right tail of the F distribution. This tells us the likelihood of getting the F statistic we actually got or one bigger, assuming the claim is true.

If we use the standard 5% level of statistical significance, we reject the claim if the $p$-value is below 5%.

*STEP* In cell F13 enter the formula: *=IF(F12<0.05,1,0)*. Run a simulation of cell F13.

You will find that it is quite rare, a few times out of a 1,000, that we do not reject the claim. This is good news because the claim is, in fact, false. The slope parameters in the DGP are not both zero.

What if the claim is true?

*STEP* Set cells A2, A3, and A4 to zero. Press F9 a few times and then run a simulation of cell F13.

As advertised, when the claim is true, we mistakenly reject the claim about 1 in 20 times.

## 26.2 Whole Model F on SNAP Benefits

The same steps can be applied to the Whole Model F test in the SNAP benefits example. Here is a short version of the steps.

1. Run the restricted regression to get the SSR with no explanatory variables.

2. Compute the F statistic and confirm it is identical to LINEST's F statistic for the unrestricted model.

3. Compute the $p$-value for the F statistic. Interpret it.

4. Use an IF statement to display a 1 when the claim is rejected. Run a simulation of this cell. What do you find?

5. Make the claim true and simulate again. What do you find this time?

## 26.3 Takeaways

Remember this: A joint hypothesis test has many flavors and applications. All are based on the idea that a restriction degrades the performance of the fitted model, as measured by the SSR.

The question then becomes: is the loss of fit severe enough that we reject the restricted model?

Determining the number of constraints can be tricky because there are many versions of the F test and ways to compare a restricted to an unrestricted regression.

One way to figure out the number of constraints is to "count the equal signs." Let's see how that works for the three F tests we have analyzed:

1. slope Exp = slope $\text{Exp}^2 = 0 \rightarrow 2$ equal signs
2. slope SNAP = slope Cash $\rightarrow$ 1 equal sign
3. slope Educ = slope Exp = slope $\text{Exp}^2 = 0 \rightarrow 3$ equal signs

It really is "the number of constraints that are being imposed by the claim." Counting the equal signs is a way to determine the number of constraints.

The Whole Model F test is a common regression diagnostic. The claim is that the entire regression is worthless. Using no explanatory variables would do as well.

The Whole Model F test is a low bar. It is quite easy to pass this test. If a regression fails this test, something is seriously wrong.

LINEST reports the Whole Model F statistic in row 4, column 1 of the output. We have to use *=F.DIST.RT(F statistic,df numerator,df denominator)* to get the $p$-value. It would have been better if LINEST reported the $p$-value instead of the F statistic.

# 27. Heteroskedasticity and the Robust SE

Heteroskedasticity (or heteroscedasticity) is from the Greek for "different" and "scatter." It is a violation of the independently, identically distributed errors requirement. Instead of an error term drawn from the same distribution (homoskedasticity), each observation has its own error distribution.

Our goal is to understand three implications of heteroskedasticity:

1. The OLS estimator of $\beta_k$, $b_k$, remains unbiased.

2. The OLS estimated SE of $b_k$, however, is broken—it is biased (typically too small) and inconsistent.

3. OLS is no longer BLUE—there is a better estimator than OLS called *Generalized Least Squares, GLS.*

After we understand what heteroskedasticity does, a second goal is to tackle the problem of what to do about it. We have two options:

1. Live with it: continue using the OLS slope estimator, but fix the broken SE by using the *robust SE.*

2. Correct it: run a regression on data that have been transformed to remove the heteroskedasticity—this is GLS.

There are many kinds of heteroskedasticity. We implement a DGP with a particular type of heteroskedasticity in Excel and use simulation to see its effects. Along the way, we will use a new Excel array function, OLSREG, in a new add-in. Unlike LINEST, it will give us robust SEs.

## 27.1 A DGP with Heteroskedasticity

There are many forms of heteroskedasticity. Here is an example of a DGP with a parameter, $\alpha$ (alpha), that controls heteroskedasticity:

$$y_i = \beta_0 + \beta_1 x_i + \epsilon_i x_i^\alpha, i = 1, ..., n$$

The $\epsilon_i$ are distributed, as usual, N(0,$\sigma$), but the $x_i^\alpha$ term will produce heteroskedasticity in $y$ for any $\alpha \neq 0$ since only zero will make $x_i^\alpha = 1$. We implement this DGP in Excel to understand the role of $\alpha$ and how it affects the OLS estimator.

*STEP* Create a sheet based on Figure 27.1. Name the cells in range A1:A4 with their labels. For $x$, create a series from 1 to 100 by 1. Epsilon is N(0,sigma) and cell C7 has the formula: *=beta0+beta1*A7+B7*A7^alpha*. Fill it down. Name the sheet *DGP*.

| | A | B | C | D |
|---|---|---|---|---|
| 1 | 9 | beta0 | | |
| 2 | 18 | beta1 | | |
| 3 | 20 | sigma | | |
| 4 | 0 | alpha | | |
| 5 | | =beta0+beta1*A7+B7*A7^alpha | | |
| 6 | x | epsilon | y | |
| 7 | 1 | -24.8420295 | 2.157970504 | |
| 8 | 2 | 3.402868231 | 48.40286823 | |
| 9 | 3 | 1.325907765 | 64.32590777 | |

Figure 27.1: Heteroskedasticity Setup.

*STEP* Make a scatter plot of $y$ as a function of $x$.

Your graph shows a tight concentration of points. Fitting a line to this cloud by eye would be easy. Notice how the dispersion of dots is the same all along the scatter plot.

But what if we change $\alpha$? Right now it is zero so the error term is clean, composed only of $\epsilon$. After all, any number raised to the zero power is 1 so the $x_i^\alpha$ term has no effect.

*STEP* Change cell A4 to 0.2 and press F9 a few times.

We have induced a little bit of heteroskedasticity. That is important for understanding what we are dealing with—heteroskedasticity is not either-or, but a continuum from weak to strong.

*STEP* Change cell A4 to 0.5, 1, and 2. Press F9 a few times after each change and look at the effect $\alpha$ has on the chart.

With 2, the heteroskedasticity is massive. Observations with low values of $x$ have almost no discernible bounce in $y$ compared to those at the right of the chart. This is intense heteroskedasticity.

## 27.2 The Effects of Heteroskedasticity

STEP Return cell A4 to 0 and use LINEST to regress $y$ on $x$ in cell range E7:F11. Press F9 a few times.

With homoskedastic errors (the word *identically* in iid), the regression is performing fabulously. The slope coefficient is quite close to its parameter value and the bounce is quite small. This means it is being accurately and precisely estimated.

The RMSE is small and so is the SSR compared to the regression (or explained) sum of squares. This is why $R^2$ is so high. The Whole Model F statistic is huge (we could compute the $p$-value, but it would be minuscule).

STEP Sim $b_1$ and its SE with 1,000 repetitions. What do you find?

Everything is working as advertised. The OLS estimator is unbiased and the average of the SEs is close to the approximate SE (the SD of the estimated slopes). We could compute the exact SE, but this is unnecessary since we have the approximate SE.

These excellent results are produced with $\alpha = 0$, which means we have homoskedasticity. What happens to the OLS regression in the presence of heteroskedasticity?

STEP Return to the *DGP* sheet and set $\alpha = 2$. Press F9 a few times.

The impact on the regression is dramatic. The slope estimate is bouncing all over the place, the RMSE and SSR have rocketed, and $R^2$ has crashed to almost zero. The F statistic is much smaller. All signs point in the same direction—this massive dose of heteroskedasticity has destroyed our excellent OLS regression results. What about the the OLS slope estimator, $b_1$?

STEP Sim $b_1$ and its SE with 1,000 repetitions. What do you find?

We need to be careful here. There is a great deal of bounce in $b_1$. It might be "far" from 18 (the parameter value), but that is because the simulation has a lot of noise in it.

Figure 27.2 shows the results with 100,000 repetitions (it took just over 6 minutes). We have evidence that $b_1$ is an unbiased estimator of $\beta_1$ since 18.2 is close to 18.

It is, in fact, true that the OLS slope estimator remains unbiased in the presence of heteroskedasticity. This is a rather surprising result and good news for the OLS estimator.

| DGP!$E$7 | | DGP!$E$8 | | Notes |
|---|---|---|---|---|
| Average | 18.200 | Average | 310.718 | alpha = 2 |
| SD | 393.2535 | SD | 36.7098 | 1) b1 is unbiased |
| Max | 1791.598 | Max | 492.924 | 2) OLS SE is biased |
| Min | -1937.815 | Min | 178.967 | |

Figure 27.2: Simulation of $b_1$ and OLS SE with $\alpha = 2$

Figure 27.2, however, also points to distressing news. The arrow highlights a comparison of the average of the estimated SEs, a little over 300, to the approximate SE of almost 400. This discrepancy suggests that the estimated SE is a biased estimator of the exact SE.

The bias does not diminish as the sample size increases. The result is clear: heteroskedasticity breaks the OLS estimated SE. We cannot rely on it and, if we do, we will produce incorrect confidence intervals and hypotheses tests.

The intuitive explanation for the bias is that the OLS estimated SE uses the RMSE to estimate $\sigma$ in the formula for the exact SE, $\frac{\sigma}{\sqrt{n}SD_x}$. But this is not the correct formula for the exact SE in the presence of heteroskedasticity. There is no single number for the spread of the errors when we have heteroskedasticity. Each observation has its own $\sigma$. Thus, the formula for the exact SE with heteroskedasticity is much more complicated.

As a general rule, the consequence of using the wrong formula for the exact SE is that heteroskedasticity produces OLS SEs that are too small. This makes us believe we are estimating the parameter more precisely than we actually are. Anything that uses these broken OLS SEs, like a confidence interval or hypothesis test, is compromised.

## 27.3 Coping with Heteroskedasticity

So we have heteroskedasticity and we know our OLS estimator is unbiased (surprising and welcome news), but the OLS SE is broken. What now?

We have two options: continue using OLS, but get the SE right or dump OLS for a better estimator. The former is easier and is the common approach.

We will not go into detail on the computation that results in what is known as the robust SE. Suffice it to say that it does not use the RMSE, a single value, as an estimate of $\sigma$ for every observation. Instead, each residual is used to estimate a spread of the error distribution for each observation.

There is no native Excel function that can compute a robust SE, but there is an Excel add-in with this functionality. It is freely available at dub.sh/regcourse.

*STEP* Download and install the OLS Regression add-in via the *Add-ins Manager* (alt, t, i keyboard shortcut).

You can access it by clicking on *OLS Regression* in the *Add-ins* tab, but we will not do this because that is suited for dead data. We have live, bouncing data with which we can show how the robust SE performs by using the OLSREG array function.

Designed to work like LINEST, OLSREG has four arguments: $x$ data, $y$ data, 0/1 for an intercept term, 0/1 for statistics. It uses labels to be more user-friendly so it requires preselecting a cell range that is $5 + k$ rows by 4 columns.

*STEP* Select cell range E13:H19, click in the formula bar, and enter the formula: *=OLSREG(A6:A106,C6:C106,1,1)*. Notice that, unlike LINEST, we include the labels for the $x$ (first argument) and $y$ (second argument) data. Unlike LINEST, the explanatory ($x$) variables are first, followed by the dependent ($y$) variable. Press ctrl-shift-enter and then F9 a few times.

Like LINEST, it updates when the sheet is recalculated. It replicates everything LINEST displays except for the degrees of freedom (which is $n - k$). Instead of the regression sum of squares, OLSREG displays the total sum of squares (TSS). Of course, RSS + SSR = TSS.

In addition to a labeled display, the OLSREG function gives us the usual OLS SE and the robust SE, labeled *RSE (HC1)*. There are several ways to compute the robust SE and HC1 is the most common. This provides us with our first option for dealing with heteroskedasticity.

*STEP* Run a simulation with 1,000 repetitions of $b_1$ and the robust SE. What do you find?

Once again, the results are quite impressive. The average of the 1,000 robust SEs is quite close to the approximate SE. This is great news! We have an estimator of the exact SE of the OLS slope estimator that we can trust.

Let's review. The OLS slope estimator in the presence of a great deal of heteroskedasticity ($\alpha = 2$) is unbiased, but its SE is biased and inconsistent. Thus, we abandon the OLS estimated SE of $b_1$ in favor of an estimator of the SE of $b_1$ that is computed differently. It is called the robust SE.

Simulation shows that this approach works because the average of 1,000 robust SEs agrees with the approximate SE of the OLS estimator of $b_1$.

## 27.4 Takeaways

Heteroskedasticity occurs when the error term is not identically distributed. When there is a pattern in the errors, such as the dispersion getting bigger, this is heteroskedasticity.

There are many forms of heteroskedasticity and there are statistical tests for it. They use the residuals to see if there is a pattern.

Heteroskedasticity is on a continuum so there can be a little or a lot of it. The greater the different spread in the errors, the bigger the impact on the OLS estimated SE.

This chapter focused on the most common response to heteroskedasticity: continue using OLS, but report robust SEs instead of the usual OLS estimated SEs.

*EXCEL TIP* Use the OLS Regression add-in to get robust SEs in Excel. It also gives much more nicely formatted regression output than LINEST.

# 28. Heteroskedasticity and GLS

The second approach to dealing with heteroskedasticity is much more aggressive. We abandon OLS entirely and use a new estimator of $b_1$ that is BLUE. We do this by removing the heteroskedasticity from the DGP via an appropriate data transformation.

The new estimator is called Generalized Least Squares (GLS) and there is clear good and bad news: the good news is that GLS beats OLS, but the bad news is that we almost never have the information needed to actually use GLS.

## 28.1 Transforming the Data

Recall that the DGP can be written as:

$$y_i = \beta_0 + \beta_1 x_i + \epsilon_i x_i^\alpha, i = 1, ..., n$$

Dividing both sides $x_i^\alpha$ gives us a clean error term:

$$\frac{y_i}{x_i^\alpha} = \frac{\beta_0}{x_i^\alpha} + \frac{\beta_1 x_i}{x_i^\alpha} + \epsilon_i, i = 1, ..., n$$

*STEP* In cell J6 of the *DGP* sheet enter the label "transformed intercept" and in cell J7, enter the formula: *=1/A7^alpha*. Fill it down. In cell K6 of the *DGP* sheet enter the label "transformed x" and in cell K7, enter the formula: *=A7/A7^alpha*. Fill it down. In cell L6 of the *DGP* sheet enter the label "transformed y" and in cell L7, enter the formula: *=C7/A7^alpha*. Fill it down.

It is critical that we know how to transform the data. In this case, we know the type of heteroskedasticity since we cooked the hypothetical data. Therefore, it is easy to see that we just need to divide by $x_i^\alpha$.

Of course, in the real world, we typically would not have this information and we would not know how to transform the data.

## 28.2 Getting GLS Estimates

With the transformed data, we can once again use OLS. Running OLS on appropriately transformed data is GLS.

We can once again rely on the Gauss-Markov Theorem, secure in the knowledge that OLS (on the transformed data) is BLUE (the best linear unbiased estimator).

We can do this with LINEST, but we have to use the no-intercept version. Although a trivial rewriting, expressing the transformed data in the usual form for $\beta_0$ and $\beta_1$ might help explain what we are doing:

$$\frac{y_i}{x_i^{\alpha}} = 0 + \beta_0 \frac{1}{x_i^{\alpha}} + \beta_1 \frac{x_i}{x_i^{\alpha}} + \epsilon_i, i = 1, ..., n$$

This equation shows that we are treating the regression as having two $x$ variables and no intercept. We make this explicit by including the zero in the equation.

We still have a linear regression because it is linear in the parameters, $\beta_0$ and $\beta_1$. This means we can use LINEST as usual to run this regression.

*STEP* Select cell range N7:P11, click in the formula bar, and enter the formula: *=LINEST(L7:L106,J7:K106,0,1)*. Press F9 a few times.

Running OLS on the appropriately transformed data is a manual labor method of doing GLS, akin to adding numbers by hand. Cell N7 is the GLS slope estimator. How does it do compared to the OLS slope estimator? We run a simulation to find out.

*STEP* Run a simulation with 10,000 repetitions of cells E7 and N7.

Both estimators are unbiased, but the GLS estimator is much more precise than OLS. The SD of 10,000 OLS slopes is about ten times bigger than that of 10,000 GLS slopes. GLS is vastly superior to OLS in this example.

This leads to an obvious question: Why don't we use GLS when we have heteroskedasticity? It is BLUE and, thus, is guaranteed to give us more precise estimates than OLS. Why settle for OLS when we can have GLS?

It is true that GLS is superior to OLS, but remember the problem with GLS: we have to know the form of the heteroskedasticity. The example used here is one of many possible varieties of heteroskedasticity. In general, we do not know the type of heteroskedasticity in the data.

Even if we knew that the correct functional form of the heteroskedasticity was $\epsilon_i x_i^\alpha$, we would also have to know the value of $\alpha$. If we do not know its value and, therefore, we estimate $\alpha$, this is called *Feasible Generalized Least Squares*, FGLS. This is beyond the scope of this book.

In the real world, we can often see and test for heteroskedasticity, but that does not tell us its form. Without that knowledge, we cannot transform the data and run a GLS regression.

Absent clear information on the type of heteroskedasticity, we must content ourselves with the first solution, continue using OLS, but replace the biased OLS estimator of the SE with the robust SE.

## 28.3 Takeaways

There are three important implications of heteroskedasticity:

1. The OLS estimator of $\beta_k$, $b_k$, remains unbiased.
2. The OLS estimated SE of $b_k$, however, is broken—it is biased (typically too small) and inconsistent.
3. OLS is no longer BLUE—there is a better estimator than OLS called GLS, but you have to know the type of heteroskedasticity to apply it.

Figure 28.1 offers a visual way to represent these three implications. First, the True OLS sampling distribution is centered on the parameter $\beta$. This shows it is unbiased.

Second, the dashed, imaginary OLS sampling distribution does not actually exist. This is what the usual OLS estimator of the SE would have us believe is true, but it is not. The robust SE is a good (biased, but consistent) estimator of the exact SE of True OLS (the sampling distribution of $b_1$).

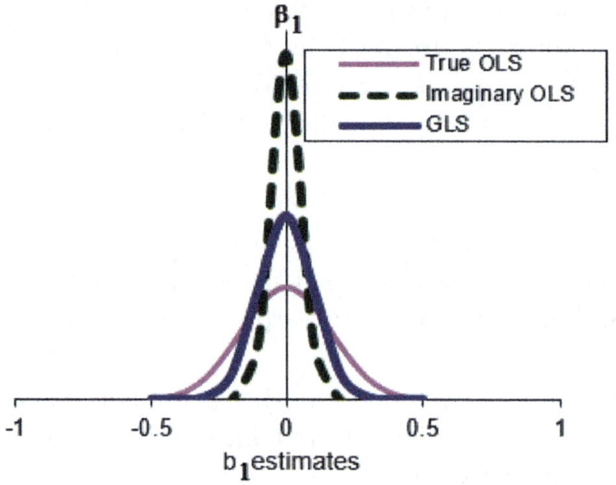


Figure 28.1: The Implications of Heteroskedasticity

Third, the GLS sampling distribution is also unbiased, but more spiked than the True OLS sampling distribution. This demonstrates the superiority of GLS.

In terms of practical econometrics, we operate by using GLS when we know the form of the heteroskedasticity, but this is not common.

The robust SE has become so popular—it was discovered only a few decades ago—that it is the default choice for many economists.

Students sometimes confuse the robust SE and think it is an alternative to OLS. It is not in the sense that we continue using the OLS recipe (choose coefficients to minimize the sum of squared residuals), but we employ a different computation for the estimated SE of those coefficients.

GLS, on the other hand, really is a different estimator of the coefficients than OLS and it produces different numerical values of the coefficients than OLS. Of course, it also produces different estimates of its exact SE.

By the way, no one would run GLS the way we did, manually transforming the data and then running OLS. Modern statistical software would do all of that for you with a single command. We did it manually to understand GLS and because Excel is not a statistical software package.

# 29. Final

*STEP* Visit dub.sh/regcourse to see what a final exam looks like for this course. Practice by taking the exam, guess how you did, then check your guess by grading yourself with the answer key. How did you do?

# 30. The Partial Xs Survey Model

We know the Classical Econometric Model (CEM) has omitted variables as part of the error term and requires fixed $X$ variables in repeated sampling. When these requirements are met, many good properties follow:

- The OLS estimator is BLUE. Not only is OLS unbiased, but it has the smallest SE (most precise) of all linear, unbiased estimators.

- The OLS estimated SE is a biased, but consistent estimator of the exact SE. We can rely on the estimated SE to give us correct confidence intervals and use it to perform tests of significance.

- Other statistics and procedures, such as the Whole Model $F$ test, work as advertised.

Conversely, if the requirements of the CEM DGP do not hold—known as violations of the DGP requirements—then the desirable properties above do not follow. Omitted variable bias and heteroskedasticity are examples of this.

But what about a different DGP than the CEM? Are there other DGPs that can be used to support linear regression and have the statistics work as advertised? Yes, and the *Partial Xs Survey Model* is one of them.

As the name implies, the Partial Xs Survey Model is a DGP that produces data by sampling from a population, this is the *survey* part of its name. But it does so without getting all of the $X$ variables that influence $Y$. This is what *Partial Xs* means.

The idea is that the missing $X$ variables are not included in the sample because they cannot be directly measured or they are outside the scope of that particular survey. As long as those omitted $X$ variables are unconnected with the included $X$ variables, all is well.

For example, say we use the Current Population Survey (CPS) for an earnings function regression. It will be missing key independent variables that affect earnings. While we do have education, experience, and demographic characteristics, the CPS does not ask questions about intelligence, ambition, and many other variables that influence earnings. These missing Xs are included in the error term of the Partial Xs Survey Model. They must be independent of the Xs included in the regression for the usual statistical procedures to work properly.

Thus, like the CEM, omitted variables play a crucial role in the error term, along with the usual "inherent randomness in the world" that is captured in the $\epsilon$ term. We repeat a crucial requirement: these omitted variables must not be correlated with the included $X$ variables. This is a difficult requirement to meet and in the case of an earnings function, it is not met.

Our goal is to implement the Partial Xs Survey Model in Excel and show that it also shares the desirable properties above. We also repeat the same core worry about the DGP—omitted variables that are correlated with included $X$ variables create omitted variable bias.

## 30.1 Partial Xs Survey Model

A multivariate version of the Partial Xs Survey Model can be written like this:

$$y_i = \beta_0 + \beta_1 x_{1i} + ... + \beta_k x_{ki} + \beta_{k+1} x_{k+1,i} + ... + \beta_{k+j} x_{k+j,i} + \epsilon_i$$

$$i = 1, ..., N \text{ indexes observations; } j = 1, ...m \text{ indexes omitted Xs in sample}$$

$$E[\epsilon_i] = 0, i = 1, ..., N$$

$$SD[\epsilon_i] = \sigma, i = 1, ..., N$$

$$\epsilon_i \text{ is distributed independently of } \epsilon_i$$

$$\text{omitted } x_{k+j,i} \text{ independent of included } x_{ki}$$

$$\text{Draw a random sample of } y_i, x_{1i}, ..., x_{ki}, i = 1, ..., n$$

The first equation says that the $y_i$ are produced by many independent variables, but it splits them into those that will be sampled (from $x_1$ to $x_k$) and those are not included in the sample (from $x_{k+1}$ to $x_{k+j}$).

Thus, like the CEM, there is a mechanism that is producing the dependent variable and it includes explanatory variables and random error ($\epsilon$). The random error term must be well-behaved. It must be independently, identically distributed (iid).

The penultimate statement in the formal model reminds us that the $X$ variables not included in the sample cannot be connected with the sampled $X$ variables. If they are, this is a violation of the DGP and we get omitted variable bias in our estimated $b_k$ slopes.

The Partial Xs Survey Model gives us a sample with $n$ observations from the population of size $N$. The population produces $Y$ based on all of the $X$ variables, but then we get a sample of $Y$ with only a subset of the $X$ variables.

The Partial Xs Survey Model has error from *epsilon* and from sampling. The sampling can be simple random sampling with replacement or something more complicated.

## 30.2 Implementing the Partial Xs Model

We return to the skiing example, with training and talent, but we simply have a population of skiers from which we sample.

The population DGP is given by this equation: This DGP can be written as a formal equation:

$$ObservedTime_i = \beta_0 + \beta_1 Talent_i + \beta_2 Training_i + \epsilon_i$$

Training will be sampled and Talent is the omitted variable. To use the SAMPLE function, we need Training to be after Talent in the spreadsheet so we switch the order of the two variables. The $\epsilon$ term is, of course, iid.

Figure 30.1 shows the initial parameter values. Notice that the correlation between Talent and Training is set to zero. This means that more talented skiers do not naturally gravitate to skiing and train more. This is unrealistic, but required by the DGP.

| | A | B | C | D |
|---|---|---|---|---|
| 1 | 80 | beta0 | (secs) | |
| 2 | -0.2 | beta1 | (secs/index points) | |
| 3 | -0.5 | beta2 | (secs/hours per day) | |
| 4 | 1 | sigma | (secs) | |
| 5 | 0 | r | | |
| 6 | 0 | mean Talent | (index points) | |
| 7 | 4 | SD Talent | (index points) | |
| 8 | 3 | mean Training | (hours per day) | |
| 9 | 2 | SD Training | (hours per day) | |

Figure 30.1: Partial Xs Survey Model Setup.

*STEP* Enter the parameter values and labels in a blank sheet and name the values with their labels. In cells A11, B11, C11, and D11 enter the labels "epsilon," "Talent," "Training," and "Time." In cell A12 enter the formula *=NORMALRANDOM(0,sigma)* and fill it down to cell A1011. There are 1,000 observations in the population.

Next, we use the BIVARNORMAL array function to draw Talent and Training at the same time.

*STEP* Select range B12:C12, click in the formula bar, enter the formula *=BIVARNORMAL(mean_Talent,SD_Talent,mean_Training,SD_Training,r_)*, and press ctrl-shift-enter. Fill it down.

Our final step in the DGP is to compute the dependent variable.

*STEP* In cell D12 enter the formula *=beta0+beta1*B12+beta2*C12+A12* and fill it down.

Before we can sample from our population, we need to give the probabilities of selection. With a simple random sample (SRS), we have equal probabilities of selection.

*STEP* In cell F11 enter the label "prob selection" and in cell F12 enter the value 0.001. Fill it down.

We can now use the SAMPLE array function to draw from the population. Our sample size will be 50 observations from the population of 1,000 and we will sample with replacement.

*STEP* In cells H11, I11, and J11 enter the labels "prob selection," "Training," and "Time." Select range H12:J61, click in the formula bar, enter the formula *=sample(C12:D1011,F12:F1011,50,1)*, and press ctrl-shirt-enter. Press F9 a few times.

You are looking directly at the Partial Xs Survey Model. On the left is the population and on the right the sample drawn from it. The sample is missing a variable, Talent, that plays a role in determining Time. Critically, however, the correlation between Talent and Time has been set to zero.

It is not, however, exactly zero in the data in columns B and C.

*STEP* In cell C10 enter the formula: *=CORREL(B12:B1011,C12:C1011)*. Press F9 a few times.

With 1,000 observations, the correlation is close to, but not exactly zero in each realization of the chance process. In the long run, as we run the chance process over and over, it is zero.

## 30.3 Properties of the Partial Xs Survey Model

How does a regression of Time on Training perform? Can it produce a good estimator—both accurate and precise—of $\beta_2$, the effect of Training on Time?

*STEP* Use LINEST to regress Time on Training with the sample data. Press F9 a few times.

The initial results look good. The estimated slope seems to be around $-0.5$, the parameter value.

*STEP* Run a simulation that tracks both the estimated slope and its SE. What do you find?

This simulation suggests that the OLS slope is an unbiased estimator of its parameter value because the average of the slopes in the simulation is close to the parameter value.

In addition, the average of the OLS estimated SEs is close to the SD of the estimated slopes. This is excellent news.

The Partial Xs Survey Model allows us to use the statistical formulas and tests that also apply to the CEM. This is a welcome result.

These rosy results and conclusions come with one crucial caveat: the omitted *Xs* must be independent of the included (sampled) *Xs*.

*STEP* Change the correlation in cell A5 to 0.9. Press F9 a few times.

The damage is immediately obvious. The estimated slope coefficient is not close to $-0.5$.

*STEP* Run a simulation that tracks both the estimated slope and its SE. What do you find?

This time the results are terrible. The OLS slope is badly biased. We think Training is much more important than it is because the highly-correlated Talent variable is omitted from the regression. Those with more Training are also likely to have more Talent so Training gets too much credit. This is classic confounding, also known as omitted variable bias.

## 30.4 Live and Dead Xs

The Partial Xs Survey Model tells a story of how the data were generated that is based on redrawing the Xs in the population. It is as if we rewound and replayed the world so each observation has new Talent and Training.

We could instead tell a different story where the *Xs* are constant and the only source of error is from epsilon.

Does a DGP based on live versus dead Xs in the population affect the statistical properties of the regression?

We can simulate to find out. We use the strategy applied to the CEM, deadening the Xs and running Solver to get a correlation between Talent and Training that is very close to zero.

*STEP* Set cell A5 to 0 and copy the *DGP* sheet. Name the new sheet "DeadXs". Select range B12:C1011 and *Paste Special: Values* to deaden the Xs. In some versions of Excel, you may have to set the sheet to *Manual Calculation*, copy/paste, then return it to *Automatic Calculation*.

After deadening the *Xs*, the correlation in cell C10 should be close to zero, but we can make it even closer with Solver.

*STEP* Run Solver with cell C10 as the objective function, Value of 0 selected, and cell C12 as the changing cell.

Now the correlation between Talent and Training is very close to zero. This fulfills the independence of included and omitted $X$ requirement of the DGP.

*STEP* Run a simulation that tracks both the estimated slope and its SE. What do you find?

Comparing these results to the simulation results with live *Xs* shows that deadening the *Xs* has no effect on the OLS slope estimator.

We can also compare other statistics in LINEST. For example, simulation of $R^2$ with live and dead *Xs* shows no difference (both have an average around 0.4 and SD a little over 0.1). We conclude that the Partial Xs Survey Model behaves the same way with live or dead *Xs*.

This result is surprising because you would think that bouncing *Xs* would produce more variability in the sample data than dead *Xs*. But this is not so. A simple example demonstrates what is going on.

*STEP* Insert a blank sheet in your workbook and enter *=RAND()* in cells A1 to A1000. Enter *0.001* in cells B1 to B1000. Select cells C1 and D1, enter the formula =SAMPLE(A1:A1000,B1:B1000,1), and press *ctrl-shift-enter*. Press *F9* a few times.

You are drawing one value from a set of 1000 uniformly-distributed, bouncing random numbers. Next, we deaden the random numbers.

*STEP* Copy cells A1:A1000, select cell E1 and paste special, values. Select cells F1 and G1, enter the formula =SAMPLE(E1:E1000,B1:B1000,1), and press *ctrl-shift-enter*. Press *F9* a few times.

In cell G1, you are drawing one value from a set of 1000 uniformly-distributed, fixed (dead) random numbers. Do we get different results from cells D1 and G1? Let's find out.

*STEP* Sim cells D1 and G1. What do you find?

They are the same. Even though the numbers are fixed, there is still bounce in the sample because of the sampling process.

It does not matter that the random numbers are bouncing because we have so many dead random numbers that reflect variability in the population.

This is why live *Xs* in the Partial Xs Survey Model produce random samples that do no harm to conventional regression statistics like the RMSE and estimated SEs of coefficients.

## 30.5 Takeaways

If we use a survey to take a random sample, whether sophisticated and large-scale like the CPS or a basic and small one, then we can use the Partial Xs Survey Model to support our use of regression analysis.

If the DGP and sampling conform to the requirements of the Partial Xs Survey Model, OLS is BLUE and the usual statistics work as advertised.

An absolutely critical requirement is the independence of the omitted *Xs* from the included *Xs*.

If the omitted *Xs* are associated with the included *Xs*, instead of being unconnected, our OLS estimator breaks down. This is called omitted variable bias and is the Achilles heel of regression analysis.

While other requirements could also be violated (think heteroskedasticity), omitted variable bias is the most serious problem we face.

The story behind the Partial Xs Survey Model is based on rewinding and replaying the world. Each observation gets new *Xs* and $\epsilon_i$ values, producing a new $y_i$.

Perhaps surprisingly, if we hold the *Xs* constant (and maintain the strict independence condition between included and omitted variables), this has no effect on the statistical properties of the regression. The live and dead *Xs* versions of the Partial Xs Survey Model are the same. This is because the sampling process is the source of variability in the sample data so whether the *Xs* bounce or not is irrelevant.

# 31. Three Ideas

This chapter is quite different from the others. Your mission, should you choose to accept it, is to open prepared Excel files, follow the instructions, and answer the questions. The files are completely standalone so there is no other reading or software needed.

The ideas are rather advanced, but they are presented with the concrete and visual philosophy embedded throughout this book. With the concepts you have mastered thus far, you can do it.

There are three files available:

1. Orthogonal Regression: The Classical Econometric Model requires fixed Xs in repeated sampling. Relaxing this requirement gives an Errors in Variables DGP and ordinary least squares (LINEST) behaves badly. A new recipe, orthogonal regression, is comes on the scene.

2. Binary Response Model: Instead of a continuous Y variable, what happens when the Y variable can only take on values of 0 or 1? This is quite common since there are many examples of this situation. You can go to college or not, get married or not, buy a car or not, and on and on.

3. Unequal Probability of Selection: We have worked with simple random sampling, but complex surveys (like the Current Population Survey) purposely overweight selection of certain groups (like Hispanics). The implications of this are not easily explained, sometimes it causes bias in the estimated slope coefficient, but other times not. Unequal probability does affect the estimated SE.

*STEP* Visit dub.sh/regcourse to download Excel workbooks on orthogonal regression, binary response, and unequal probability of selection.

Good luck!

[illegible]

[illegible]

[illegible]

[illegible]

[illegible]

[illegible] the settlement [illegible]

[illegible]

[illegible]

# 32. The Journey Ends

We began by asking, what is the rate of return to education? We used this question to learn about regression. We saw how using a semi-log functional form captured the multiplicative nature of education as an investment. We also saw how we needed to add variables to the regression to statistically control for other factors.

We used Monte Carlo simulation to understand that sampling produces a regression slope coefficient that is a random variable with a sampling distribution. The SE is a statistic that tells us the dispersion in the estimate.

We demonstrated with simulation how the Gauss-Markov Theorem works. We learned about confidence intervals and hypothesis testing. More recently, we studied the effect of heteroskedasticity.

In this final chapter, we put all of this together to get a final estimate of the rate of return to education with the tools and concepts used in this book.

This chapter has no steps for you. It simply shows some of the important ideas we have covered and offers a model for how you would display regression results. We use Stata, but any sophisticated statistical package (for example R, Python, or SAS) would produce the same results.

## 32.1 The 2024 ASEC

From cps.ipums.org, we selected the 2024 ASEC (March CPS). We downloaded the variables INCWAGE, EDUC, AGE (25-64 years old), SEX, RACE, HISPAN, MARST, STATEFIP, UNION, CLASSWKR.

We dropped all incomes under $20,000 and created educyears by recoding the EDUC variable into number of years of schooling from 10 to 20 years. We created exp (a rough measure of experience) as AGE - 6 - educyears.

Recoding is really important when working with survey data like the CPS. Many questions are categorical, like race or state of residence. They are all numbers, however, so software will happily treat what are categories as numbers. It is easier than it should be to make a fatal error and forget to indicate that the variable is not a number (like age or income).

We used IF statements and VLOOKUP in Excel to recode categorical variables into dummy variables, but modern statistical software does this hard work for you. For example, in Stata, you simply put an i. (i dot) in front of the variable, like i.race or i.statefip and it creates all of the dummy variables.

## 32.2 Education ROR

Figure 32.1 shows that the estimated coefficient on educyears is 0.1166. In Excel, entering the formula =EXP(0.1166) - 1 gives approximately 12.4%. This means that each additional year of education is associated with about a 12.4% increase in annual earned income.

```
. reg lnincwage educyears

      Source |       SS           df       MS      Number of obs   =    50,124
-------------+----------------------------------   F(1, 50122)     =  12234.04
       Model |  4198.50276         1  4198.50276   Prob > F        =    0.0000
    Residual |   17200.977    50,122  .343182176   R-squared       =    0.1962
-------------+----------------------------------   Adj R-squared   =    0.1962
       Total |  21399.4798    50,123  .426939325   Root MSE        =    .58582

------------------------------------------------------------------------------
   lnincwage |      Coef.   Std. Err.      t    P>|t|     [95% Conf. Interval]
-------------+----------------------------------------------------------------
   educyears |   .1166089   .0010543   110.61   0.000     .1145425    .1186752
       _cons |   9.359285   .0157106   595.73   0.000     9.328492    9.390078
------------------------------------------------------------------------------
```

Figure 32.1: Regressing lnincwage on educyears.

Of course, as you know, nothing is being held constant here. There are no statistical controls. This raw differential is misleading because it seems the income of someone with 12 years of education who magically gained another year of schooling, would increase by 12.4%. This is not true.

The people with 13 years of education are different in many other ways than the people with 12 years of education. Bivariate regression does not control for any of these other differences. This is the heart of the problem with omitted variable bias.

Another important idea from the regression results is the SE (labeled Std. Err in Figure 32.1) of the educyears coefficient. It is really small relative to the slope coefficient itself. This is great news because it means the coefficient is being very precisely estimated because the sample size is really big, 50,0124 observations.

The small SE leads to tight confidence intervals. You can see that the 95% confidence interval is really small, from 0.1145425 to 0.1186752.

The small SE also drives the highly significant hypothesis test result. The $t$ statistic is simply 0.1166089 divided by 0.0010543. Stata reports the $p$-value ($P > |t|$) as 0.000, but it is not really zero. The $p$-value is, however, incredibly small, which means we can reject the null that the coefficient is actually zero.

Does highly significant mean we have the really exact, correct answer? No. It just means that our estimate does not have a lot of variability so its sampling distribution is very spiked and we can reject the claim that the truth is zero. That says nothing about whether or not it is centered on the true, unknown parameter value.

Does highly significant mean this is a practically important result? No. It just means that the chances are ridiculously small that the true value is zero and we got this sample result by luck alone.

A hypothesis test uses the null to place the center of the sampling distribution at the null value (usually zero). You use the estimated SE to draw the sampling distribution and you know the one estimate you observed. This allows you to determine the chances of getting an estimate like you actually got (or more extreme). But we do not actually know where the sampling distribution is centered because we do not know the parameter value.

Remember that statistical significance answers a specific question and only that question. It just means that chance is unlikely to have produced a sample outcome like the one we have if the claim is true. That is it. There is nothing else to say. Believing that statistical significance means exactly right or important reflects serious misunderstanding.

The regression result is, in fact, practically important in the sense that education really does exert a strong pull on income. A 12.4% increase in annual earned income per year of education is a lot of money. Consider that the average income is roughly $83,000 so 12.4% of that is about $10,000 per year. And the dollar increase rises as education goes up!

Note that the R-squared is almost 0.2, which means that education alone accounts for about 20% of the variation in income. There is no doubt about it, education is a powerful factor in determining income.

But is it really 12.4% per year of education? Maybe, but we are worried about omitted variable bias so we will estimate a second model. We will keep paying attention to the coefficient on educyears, tracking how it changes as we add variables to the regression.

Our first control variable is experience. But we model it as experience and experience squared to capture the non-linearity in the average income as experience increases.

A model specification of the natural log of some measure of earnings as a function of education, experience, and experience squared is a minimal model for an earnings function.

Figure 32.2 has the results. The slope coefficient on educyears has increased a little. The interpretation is the same as before, the percentage increase in income, on average, with each additional year of education, but we add, "holding experience constant." This is critical to correctly interpreting the coefficient.

```
. reg lnincwage educyears exp expsq

      Source |       SS           df       MS      Number of obs   =    50,124
-------------+----------------------------------   F(3, 50120)     =   4567.60
       Model |  4594.48715         3  1531.49572   Prob > F        =    0.0000
    Residual |  16804.9926    50,120  .335295145   R-squared       =    0.2147
-------------+----------------------------------   Adj R-squared   =    0.2147
       Total |  21399.4798    50,123  .426939325   Root MSE        =    .57905

------------------------------------------------------------------------------
   lnincwage |      Coef.   Std. Err.      t    P>|t|     [95% Conf. Interval]
-------------+----------------------------------------------------------------
   educyears |   .1225751   .0010755   113.97   0.000      .120467    .1246831
         exp |   .0276792   .0010152    27.27   0.000     .0256895     .029669
       expsq |  -.0004516   .0000209   -21.57   0.000    -.0004926   -.0004106
       _cons |   8.932595    .019893   449.03   0.000     8.893605    8.971586
------------------------------------------------------------------------------
```

Figure 32.2: Regressing lnincwage on educyears and experience.

We display the two different models by listing the variables and showing their coefficients and estimated SEs. Figure 32.3 shows how this conveniently enables comparison of the two model specifications.

The coefficient on education did not change by much. Sometimes, however, there are big changes and even sign reversals as more controls are added.

| | (1) lnincwage | (2) lnincwage |
|---|---|---|
| educyears | **0.1166*** ** | **0.1226*** ** |
| | **(0.0011)** | **(0.0011)** |
| exp | | **0.0277*** ** |
| | | **(0.0010)** |
| expsq | | **-0.0005*** ** |
| | | **(0.0000)** |
| R-squared | **0.1962** | **0.2147** |

Standard errors in parentheses
* p<0.10, ** p<0.05, *** p<0.01

Figure 32.3: Displaying Regression Results.

Figure 32.3 has a formatting problem. Can you spot it? The estimated SE on expsq is (0.0000). In fact, it cannot be zero. We could display more decimal places, but a common workaround is to divide expsq by 100. This shifts by two decimal places the coefficients and SEs on exp and expsq.

Let's be sure you are comfortable reading regression results in table form, as in Figure 32.3. Recall the regression equation from chapter 8:

$$\text{Predicted}lnINCWAGE = 8.9326 + 0.1226 EducYears + 0.0277 Experience \\ - 0.00045 ExperienceSquared$$

Note that the fitted regression equation always starts with *Predicted Y* and it does not have an error term, unlike the equation for the DGP.

Figure 32.3 does not display the estimated intercept (also known as the constant term), but you can see how the other coefficients in the table match the equation (with the coefficient on expsq rounded to - 0.0005 in the regression table).

The estimated SEs give you the variability of each estimate. Once again, the SEs are small relative to the coefficients in this case, which is common with extremely large sample sizes.

What happens if we add more controls and use robust SEs and probability weights? Figure 32.4 tells us that the coefficient on education remains pretty stable.

| | (1) lnincwage | (2) lnincwage | (3) lnincwage | (4) lnincwage | (5) lnincwage |
|---|---|---|---|---|---|
| educyears | 0.1166*** | 0.1226*** | 0.1167*** | 0.1167*** | 0.1184*** |
| | (0.0011) | (0.0011) | (0.0011) | (0.0012) | (0.0014) |
| R-squared | 0.1962 | 0.2147 | 0.3062 | 0.3062 | 0.2945 |

Standard errors in parentheses
(1) no controls, (2) + exp expsq, (3) full model (4) robust SE, (5) + ASEC weights.
* p<0.05, ** p<0.01, *** p<0.001

Figure 32.4: Education ROR with different specifications.

Note that the coefficient is the same for models 3 and 4. This is because both models have the same regressors. The only difference is that model 3 uses the OLS estimated SE and model 4 uses the robust SE. Model 4 is better even though the SE increases just a little bit.

Can we conclude that model 5 gives the true, unknown rate of return on education? Alas, no. We are still explaining less than a third of the variation in income. There are many factors we have not included, such as ambition and intelligence (we captured this as drive).

Thew core problem with using CPS data to estimate the rate of return to education is that we have an observational study and we know people with more drive choose more education. Since we do not have any information drive we cannot include it as a statistical control. This makes it impossible to isolate the true causal effect of schooling on income.

Also, we do not have any information on the costs of additional education so we do not have a true ROR, just a measure of how income was connected to education in 2023. This does not tell us what the future relationship between income and education will be. Since we live in such interesting and unstable times, everyone can agree that projecting future income flows for different levels of education is incredibly difficult.

Here is what we can say. Our best estimate, using the tools we have, is model 5 in Figure 32.4. It includes several controls, robust SEs, and probability weights. It is better analysis than the other models.

It is encouraging that the coefficient is essentially stable across our different specifications, but that does not mean that it would stay stable if we added more controls.

We can say that our current answer for the return to education is about 12% per year of education, with less than a tenth of a percentage point in variability. We remain extremely concerned with omitted variable bias and the future is unknowable, but it does seem that education is a good investment.

This is on the high end of the many studies of the rate of return to education. The literature is vast. Mincerian earnings functions are everywhere. Usually, the range of estimates are around 8% to 12%, but there are lower and higher estimates also.

There is one critical additional factor affecting the rate of return to education—you have to finish. Starting college and investing tens of thousands of dollars, but not graduating is a terrible investment. The market pays a little more for one, two, or three years of education beyond high school, but average income really jumps upon graduation.

Many people who do not go to college are rightfully nervous about their ability to finish. In other words, investing in education carries risk so the ROR is difficult to measure. This is yet another complication that is not included in typical earnings function regressions.

## 32.3 Takeaways

There are too many things to list in this final takeaways section, so let's focus on a few fundamental ideas in this book.

We began our study of regression by exploring the relationship between income and education. We added statistical controls (experience and demographic characteristics) in an attempt to lessen omitted variable bias.

Then we turned to inference and learned how the SE is computed and interpreted. We demonstrated the Gauss-Markov Theorem: if the DGP is well-behaved, then OLS is BLUE. This is why we use OLS (LINEST).

Heteroskedasticity biases the OLE estimated SE and the common solution is to use robust SEs.

There is no test for omitted variable bias and the only way to diagnose it is via intimate knowledge of the DGP and exceedingly sharp intellect.

Observational studies almost certainly suffer from omitted variable bias. Randomized controlled experiments are better because unobserved, latent factors are randomly assigned.

You have to know your DGP. You have to know where the numbers came from, how they were produced, and what is the source of their variability.

This chapter is called *The Journey Ends*, but does not end work on the earnings function. It never ends. There are always better data and tools and techniques to explore.

I hope this book has given you some fundamental principles and skills needed for you to continue your journey in econometrics and data analytics.

# Regression with Microdata and Microsoft Excel®

Actual URL: barretoh.github.io/homepage/courses/Reg
Shortcut: dub.sh/regcourse

This text uses Microsoft Excel® to deliver an introductory regression course. A statistics course or familiarity with basic statistical concepts is recommended. The target audience is undergraduate students who have used a spreadsheet before, but have no specialized training in Excel or computer programming.

Readers acquire sophisticated Excel skills and a solid foundation for further work in quantitative courses. With concrete examples, novel applications, and eye-catching visualizations, engagement is guaranteed as the user creates spreadsheets, manipulates live graphs, and analyzes data.

Humberto Barreto is Professor of Business Analytics at DePauw University. He earned his Ph.D. from the University of North Carolina at Chapel Hill. Professor Barreto has lectured around the world on teaching economics with computer-based methods, including Cuba, Brazil, Canada, Scotland, Spain, Poland, India, Myanmar, Japan, and Taiwan. He was a Fulbright Scholar in the Dominican Republic and has taught National Science Foundation (NSF) Chautauqua short courses using simulation. He has received several research and teaching awards. His book, *The Entrepreneur in Microeconomic Theory*, was translated into Arabic in 1999. He has written numerous articles and books on using Excel to teach economics (including introductory level material, micro, macro and econometrics). His most recent book is *Gateway to Business Analytics* (2025). Visit his website for more information: barretoh.github.io/homepage.

Made in United States
Orlando, FL
04 November 2026

90200951R00162